Glaciers and Climate Change

Glaciers and Climate Change

Johannes Oerlemans

A.A. BALKEMA PUBLISHERS LISSE / ABINGDON / EXTON (PA) / TOKYO

Library of Congress Cataloging-in-Publication Data

Oerlemans, J. (Johannes), 1950-
Glaciers and climate change / Johannes Oerlemans.
p. cm.
Includes bibliographical references (p.).
ISBN 9026518137
1. Glaciers. 2. Climate changes. I. Title.

GB2405 O47 2001
551.31'2--dc21

2001040014

Cover design: Studio Jan de Boer, Amsterdam
Printed in the Netherlands by Wilco b.v.

ISBN 90 265 1813 7 (hardback)

to Roos, Gijs and Esther

Preface

For many years now I have been planning to write a book about glaciers and their response to climate change, a subject that has fascinated me since my student days. In 1999 I was able to spend a sabbatical at Stockholm University and at last found time to turn notes and sketches into a manuscript. As every author of a textbook soon discovers, the task of collecting illustrations and making suitable graphs always takes longer than expected, and yet another year passed before the manuscript was almost in a publishable form.
Although my book focuses on valley glaciers, it is not intended to provide a basic course in glaciology, nor does it claim to give a state-of-the-art picture of glacier research. It consists mainly of the personal reflections of a meteorologist who gradually became interested in glaciers and is written primarily for persons with a general interest in the physics of the environment. At the same time, however, the book contains a wide range of source material that should prove useful in university courses. Physical processes are described in a quantitative way and where possible are illustrated with the help of relatively simple models.
Many excellent texts have been written about glaciers, often with superb illustrations (see references in Chapter 1). Although parts of my text inevitably overlap with sections of these books, I put more emphasis on the interaction between glaciers and the atmosphere. Over the past 15 years the IMAU, the Institute for Marine and Atmospheric Research at Utrecht University, in collaboration with other research institutes, has conducted a series of meteorological experiments on glaciers. These experiments have provided a wealth of information about the microclimate of glaciers, and some of the data have found their way into this text.
I have attempted to find a balance between theory and observations. In a way, this book reflects my attitude towards the natural sciences. Although I started as a 'modeller', I learned in the course of time that one cannot really judge the value and nature of observations unless one has been involved in data collection. Neither can one appreciate the importance of modelling simply by reading articles. Whenever the opportunity arises one should do both !

I should like to thank all the persons and institutions that have supported my work over the years. I have had the privilege of working with the skilled and dedicated members of the IMAU ice-and-climate group and have been very fortunate to have had such good students and post-docs. This book has benefited in so many ways from what we have learned from each other during fieldwork, seminars and discussions. In particular, I want to mention Kees van der Veen (who in the meantime has written his own impressive textbook), Wouter Greuell, Roderik van de Wal, Richard Bintanja, Michiel van den Broeke, Wouter Knap and Nicole van Lipzig.
The book contains material from many projects. Through the years financial support has come from a number of bodies, notably the Netherlands Organization for Scientific Research and the various research programmes of the European Union. On behalf of our research group I should like to thank the funding agencies for placing their continued trust in our work.
Shaping a book is a process that needs feedback, even in the final stages. In addition to the persons mentioned above, I am indebted to Andrew Macintosh, Lisette Klok, Bruce Denby, Martijn de Ruijter de Wildt and Carleen Reijmer, who made useful comments on the text.
I am grateful to Sheila McNab for making my English more readable and for improving textual clarity.
Finally, a thank-you to my colleagues and friends at the Meteorological Institute of Stockholm University for their hospitality during my sabbatical. The beautiful city of Stockholm, situated in a landscape that emerged from the last ice age not long ago, provided just the right atmosphere for writing my book. Above all, the expensive cappuccinos shared with Branko Grisogono under the old oak tree kept me in the right mood !

Johannes Oerlemans
Utrecht, The Netherlands, October 2000

CONTENTS

SOURCES AND COPYRIGHT OF FIGURES REPRODUCED IN THIS BOOK

Figures 2.1, 3.8, 3.22,10.9, 10.12, 10.13: *Journal of Glaciology* (published by the International Glaciological Society).

Figure 2.11: *Zeitschrift für Gletscherkunde und Glazialgeologie* (published by Universitätsverlag Wagner, Innsbruck, Austria).

Figure 3.13: *International Journal of Remote Sensing* (published by the Remote Sensing Society).

Figure 3.14: *Boundary-Layer Meteorology* (published by Elsevier Science).

Figure 3.19: *Journal of Applied Meteorology* (published by the American Meteorological Society).

Figure 5.1: Wideroe, Oslo, Norway.

Figures 7.9, 8.9, 8.10, 8.11: *Annals of Glaciology* (published by the International Glaciological Society, Cambridge, U.K.)

Figure 9.6: *Arctic, Antarctic, and Alpine Research* (published by the Institute of Arctic and Alpine Research, University of Colorado, U.S.A.).

Figure 10.2: *Science* (published by the American Association for the Advancement of Science).

Figure 10.11, 11.1, 11.2, 11.3, 11.4, 11.5: *Climate Dynamics* (published by Springer-Verlag, Berlin)

Figure 10.14: N. van Lipzig

Figure 12.2: C. Reijmer

Figure 12.2: M. de Ruyter de Wildt

Figure 12.5, 12.6, 12.12: J.W. Greuell

Figures 12.7, 12.8: H. Rott

Figure 12.9, 12.10: *Geophysical Research Letters* (published by the American Geophysical Union).

Figure 12.11: J.A. Dowdeswell

1. Introduction

Mountains have fascinated and inspired mankind throughout history. Mountains with glaciers are even more exciting, especially when these glaciers flow down into the valleys, into the areas where people live. Large valley glaciers look solid and very robust, but in fact they are changing all the time. The lower they come, the greater the contrast with the surrounding landscape. Many glaciers in the world enter pastures and forests. It is not surprising that laymen, artists and scientists, each in their own way, have reported on the behaviour of large valley glaciers. A wealth of information on glacier fluctuations in historical times has thus become available, notably from the Alps, Norway and Iceland. Now that this information has been combined with the results of geomorphological studies it has become possible to reconstruct the past behaviour of glaciers in some detail.

The study of glaciers is a highly rewarding academic occupation. The interest is shared by the general public: glaciers are fascinating and deserve attention, just like hurricanes or volcanoes... However, there are also more practical reasons for studying glaciers.

Glaciers have peak discharges when the need is largest: in hot summers (*figure1.1*). Consequently, in many parts of the world glacier melt water is used to supply irrigation systems. Hydropower reservoirs may also benefit from the presence of glaciers. Reservoirs are normally emptied in winter when the need for power is largest. In the summer melt water from glaciers helps to refill them. In view of this, it is not surprising that hydropower companies (particularly in Switzerland, Iceland and Norway) have supported mass balance studies over the years.

Glaciers, however, also have their negative sides. Sudden events like ice avalanches and outbursts of glacier-dammed lakes can cause catastrophes with many casualties and extensive destruction. Fluctuations of a more regular kind like alterations in glacier extent can threaten roads, constructions and property.

Figure 1.1. View towards the terminus of Nigardsbreen, Norway. This glacier has an area of about 48 km^2. On warm summer days the peak discharge of the glacial river is typically 10 $m^3\ s^{-1}$. Note the person standing at the river.

Considered on a global scale, valley glaciers and small ice caps are believed to contribute significantly to sea-level fluctuations on a century time scale. Compared to the large ice sheets of Greenland and Antarctica, the amount of water stored in glaciers and ice caps is really small (about 0.5 m of sea-level equivalent). However, glaciers are in warmer and wetter climates and therefore have a relatively larger mass turnover. This makes them more sensitive to climate change. Also response times are smaller. It is believed that the retreat of glaciers and ice caps during the last hundred years has contributed several centimetres to global sea-level rise.

Glaciers are a fascinating laboratory for the study of micro-meteorological processes. Even if one is not concerned with glacier mass balance, glacier fluctuations and climate change, it is interesting to carry out meteorological studies on glaciers because extreme conditions are often encountered. In summer, vertical temperature gradients of 10 K over 2 m accompanied by wind speeds of 5 m s^{-1} are quite common on glacier tongues. Rapid changes of the surface during melting are another example. There is a strong coupling between energy exchange, melt of snow and ice, and albedo and this makes the surface processes very dynamic on the time scale of weeks. Even when interest is mainly in the decadal-to-century time scale, such processes have to be understood.

The backbone of this book is shown in *figure 1.2.* The response of the geometry of a glacier to climate change is the outcome of a process that consists of several steps. Changes in the large-scale climate setting will be felt at the glacier surface and will thus affect the mass balance. But it is important to realise that a large-scale climatic signal is filtered by the atmospheric boundary-layer before it reaches the glacier surface. So a mass-balance model that translates weather conditions into net mass gain or loss at the glacier surface needs to contain a good deal of meteorological theory.

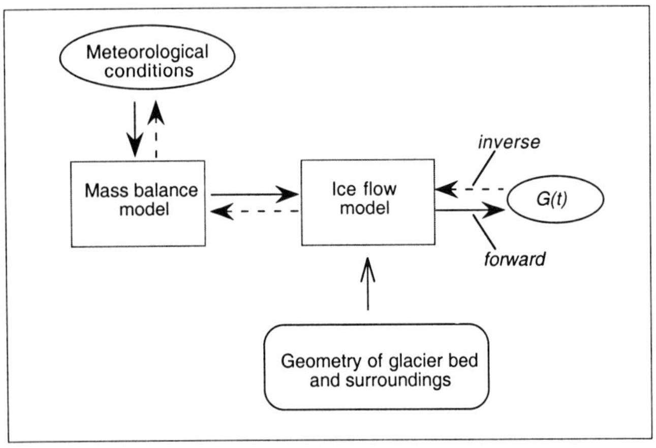

Figure 1.2. Relating the evolution of glacier geometry G(t) to meteorological conditions. In the forward mode, the mass-balance model serves as forcing for the ice-flow model. However, deducing climate change from glacier fluctuations (the inverse problem) is of great interest as well.

The calculated distribution of mass gain or loss serves as forcing for an ice-flow model, which calculates changes in glacier geometry in the course of time. This ice-flow model should be compatible with the mass-balance model. It makes little sense to combine a very detailed treatment of ice mechanics with a primitive description of the relation between climate and mass balance.

The solid arrows in *figure 1.2* mark the forward path: finding out, for instance, how glaciers will respond to greenhouse warming scenarios and contribute to sea-level rise. But the backward problem, indicated by the dashed arrows, is also important. In principle, inverse modelling makes it possible to derive a climate signal from glacier fluctuations in the past.

The book is organised as follows. In Chapter 2 basic concepts are introduced. Readers familiar with glaciers may skip this chapter and proceed to Chapter 3. In a way, Chapter 3, which deals with the microclimate of glaciers, forms the heart of the book because it describes the processes that regulate the exchange of heat and mass between atmosphere and glacier surface. The next step is to study the mass balance, and results of mass-balance modelling are used to illustrate the fundamental differences between maritime and continental glaciers (Chapter 4). Chapter 5 deals with ice mechanics. The treatment of ice mechanics is limited and rather elementary. Analytical models are introduced to describe how glaciers with a simple geometry respond to climate change.

Real glaciers cannot be modelled in detail without the help of numerical methods. A numerical one-dimensional flowline model, now used by many researchers in the field, is introduced and used to study the basic behaviour of glaciers with different geometries and climatic settings (Chapter 6, 7 and 8).

In Chapter 9 theory and results of modelling come together. Here the interpretation of glacier fluctuations is considered and the possibilities and limitations of inverse modelling are discussed. Sea-level rise is an issue that deserves attention in a book about glaciers, and the subject is embedded in Chapter 10. Here one can also find a short discussion on the state of balance of the Greenland and Antarctic ice sheets; an attempt is made to compare the contributions made to sea-level change by changes in the volume of land ice and the thermal expansion of the oceans.

Chapter 11 is about the future of glaciers. We wonder how glaciers will fare in the next hundred years. Glaciers have retreated worldwide, and it seems likely that enhanced greenhouse warming will lead to further recession. But it is not so easy to predict how quickly this will occur and what the consequences will be for global sea-level rise. Results of the glacier-modelling study carried out by EISMINT (European Ice Sheet Modelling Initiative, which was coordinated by the European Science Foundation) will be discussed.

Chapter 12 gives a brief survey of recent developments in research methods. Assessing glacier mass balance from space is a challenging task. Modern equipment now also permits the collection of in situ data in a more automated fashion. The book concludes with some thoughts about how new measuring techniques and numerical models can be combined to make glacier research more effective.

A selection of books about glaciers

1. *Die Schweiz und ihre Gletscher - Von der Eiszeit bis zur Gegenwart* (1981). Kümmerly und Frey Geographischer Verlag (Bern), 191 pp.

2. H.J. Zumbühl, B. Messerli and C. Pfister (1983). *Die kleine Eiszeit - Gletschergeschichte im Spiegel der Kunst.* Steiner AV (Sarnen, Switzerland), 60 pp.

3. W.S.B. Paterson (1994). *The Physics of Glaciers.* Third edition. Pergamon/Elsevier, New York, 380 pp.

4. G. Østrem and M. Brugman (1991): *Glacier mass-balance measurements - A manual for field and office work.* National Hydrology Research Institute, Canada, 224 pp.

5. M. Hambrey and J. Alean (1992): *Glaciers.* Cambridge University Press, 208 pp.

6. B. Wold and L. Ryvarden (1996): *Jostedalsbreen - Norway's largest glacier.* Boksenteret A/S, Oslo, 95 pp.

7. R.A. Souchez and R.D. Lorrain (1991): *Ice Composition and Glacier Dynamics.* Springer-Verlag, Berlin, 207 pp.

8. R. LeB. Hooke (1998): *Principles of Glacier Mechanics.* Prentice Hall, New Jersey, 248 pp.

9. C.J. van der Veen (1999): *Fundamentals of Glacier Dynamics.* Balkema, Rotterdam, 406 pp.

2. Basic concepts

2.1 Where do glaciers form ?

Glaciers form when the snowpack accumulated in winter (or, in the case of tropical glaciers, in the wet season) does not melt entirely in summer, at least in most years. The remaining snow then turns into ice to form a glacier. Glaciers therefore tend to be located in wet and cold places. Wet implies a region not too far from a moisture source (the ocean) located in upwind direction; cold implies high latitude or altitude. Temperature and precipitation are not entirely independent factors. If we travel polewards the mean precipitation becomes less and less. This is simply because the moisture-holding capacity of the air decreases strongly when the temperature drops. As a result, the windward side of the high mountain ranges located in the westerlies is the best place to find glaciers. Indeed this is where the most active glaciers are to be found, for example in Norway, Alaska, Patagonia and New Zealand.

Most mountain ranges show a marked asymmetry in the precipitation regime. The asymmetry is of course greater when the atmospheric circulation is more persistent. Extreme differences are for example found in New Zealand, but the Scandinavian mountains show this feature too. Western Norway receives several metres of precipitation in a year, whereas some regions in eastern Norway and Sweden receive only half a metre. The Alps present another picture. In the course of the year, storms move in from all directions. Here we find a pattern of high precipitation (~2 m) in the outer mountain ranges and somewhat drier conditions in the interior.
Outside Greenland and Antarctica there are tens of thousands of glaciers, many of them really small (less than a few square kilometres). They are very unevenly distributed over the globe. *Figure 2.1* has been constructed with data from the World Glacier Monitoring Service (Zürich). It shows the glacier area for a hundred regions where glaciers are found.

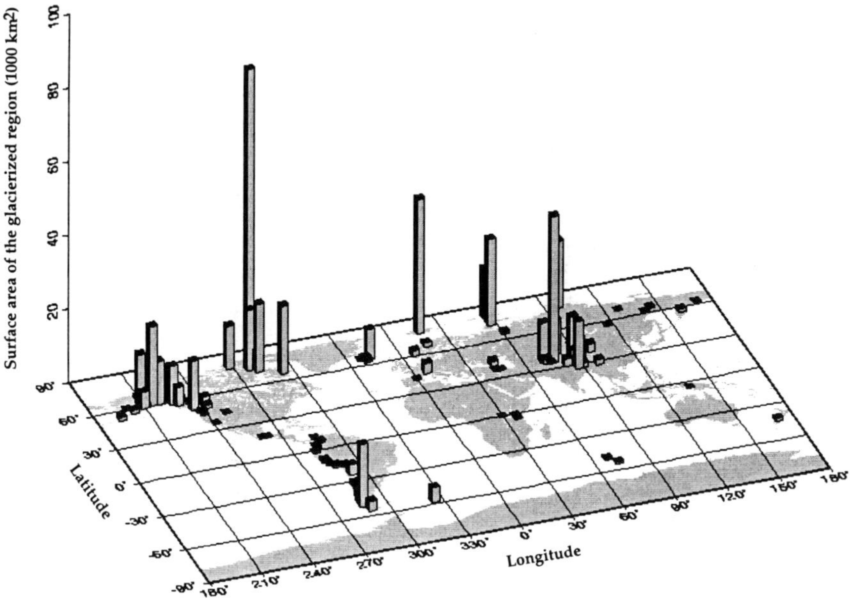

Figure 2.1. The area of a hundred glacierised regions in the world outside Greenland and Antarctica. Prepared by Z. Zuo from data collected by the World Glacier Monitoring Service [1,2]. The hundred regions are listed in Chapter 10, *Table 10.2.*

The glacier regions include valley glaciers, ice fields and ice caps. The definitions of these terms are not very strict. An ice cap normally covers a plateau-like terrain. It can have smaller or larger outlet glaciers that flow onto lower land and eventually into the sea. A fine example of a large ice cap is Vatnajökull on Iceland. It has an area of about 8000 km^2. A large part of the ice flows out in wide lobes, a smaller part flows into valleys to form valley glaciers. When valley glaciers are very large and are connected to cover a large area the term ice field is frequently used. Examples are the ice fields of North Patagonia and South Patagonia and Alaska.
The total glacier area outside Greenland and Antarctica is estimated to be 530 000 km^2. When the glaciers and small ice caps on Greenland and Antarctica (not being part of the major ice sheets) are included, the area is 680 000 km^2 [3].

2.2 Specific balance and equilibrium line

Snowfall during an intense storm, typically lasting a few days, is the most obvious process that adds mass to a glacier. The counterpart is melt and runoff on a warm summer day. The net result of these and other processes after one year, expressed in mass gain (+) or loss (-), is called the specific balance. The specific balance rate, denoted in this book by B, is expressed in kg m^{-2} a^{-1}, or in mwe a^{-1} (metres of water equivalent per year). In the literature, the terms balance (mwe) and balance rate (mwe a^{-1}) are not always used in a consistent way. However, in equations it is normally clear what is meant.

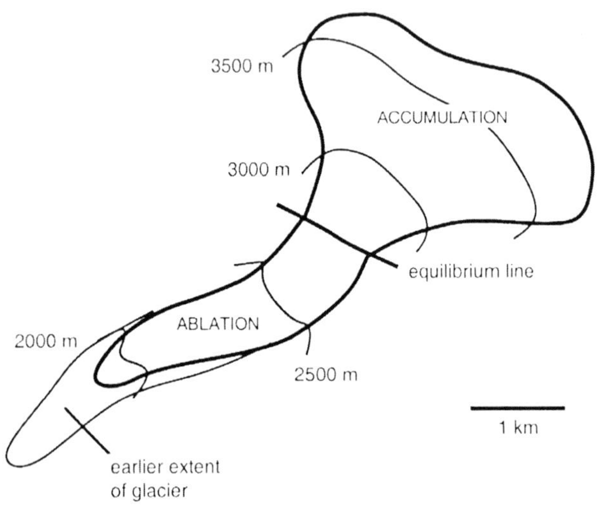

Figure 2.2. Schematic picture of a valley glacier. The accumulation and ablation areas are separated by the equilibrium line, at which, by definition, the specific balance is zero. On most glaciers the accumulation area is larger than the ablation area. Climatic change implies a shift of the equilibrium line.

B can be measured in several ways, as is explained later. Observations made on many glaciers show the same basic pattern. On the lowest part of a glacier B is strongly negative. For instance, on the snout of Nigardsbreen (*figure 1.1*), Norwegian observers have measured values of typically -10 mwe a^{-1}. Further up, the specific balance becomes less negative and ultimately positive.
Since altitude is the most important factor, it makes sense to define an equilibrium line and the corresponding equilibrium-line altitude (frequently abbreviated to ELA, in this book denoted by E). The equilibrium line connects all the points on a glacier which have zero specific balance.
The equilibrium-line concept is only useful if some smoothing of the mass balance field is allowed for. On all glaciers there is significant spatial variability in snow accumulation and melt. As a consequence the specific balance does not always follow a simple pattern. In fact, the equilibrium line can form closed loops and does not necessarily run smoothly from one side of the glacier to the other. However, when smoothing is applied, the equilibium line on larger glaciers will more or less follow an isohyps. In most cases this means that the equilibrium line does run smoothly from one side of the glacier to the other, separating the accumulation zone ($B > 0$) from the ablation zone ($B < 0$). The resulting picture is summarised in *figure 2.2*.
There are other 'lines', 'limits' and 'zones' on a glacier. Above the dry snow line, for instance, melt does not occur. The percolation zone is a zone above the equilibrium line where in summer melt water forms and percolates into the snowpack, but does not run off. In the superimposed ice zone melt water refreezes in late spring/early summer. The layers formed do not melt away entirely later in the ablation season. The classification of zones, see [7], is discussed briefly in the next section.
From a more general point of view, the climatic conditions can be represented by an equilibrium surface, which gives for any location the altitude at which a fictitious glacier surface would have zero specific balance. The equilibrium line on a real glacier then is where the equilibrium plane intersects with the glacier surface. When climatic changes occur the equilibrium surface will go up and down.
Without referring explicitly to glacier dynamics we can formulate a condition that should be fulfilled for a glacier in steady state. The mean mass balance B_m of a mapped glacier is defined as

$$B_m = \int_{area} B \, dx \, dy \, . \qquad (2.1)$$

Here dx and dy are the infinitesimal increments along the map coordinates. The total mass of a glacier will not change if

$$B_m = 0 \, . \qquad (2.2)$$

The melt underneath a glacier has been neglected (it is no

more than a few mm of ice per year).

One sometimes comes across the statement that the equilibrium surface more or less follows annual or summer temperature. We now know that this is a poor approximation of the real situation. Even if the rate of melting were perfectly related to air temperature, differences in snow accumulation would lead to differences in E. For instance, more snowfall would imply that a higher air temperature would be needed to melt the snowpack and the equilibrium surface would be lower. In addition, many meteorological studies on glaciers have shown that most of the melt is caused directly by solar radiation. Because the amount of solar radiation impinging on the earth surface varies strongly from place to place, an even larger variability in E has to be expected. A proper study of the relation between climate and E therefore requires full consideration of the energy budget.

The equilibrium line is a mathematical concept that cannot be observed directly. However, when looking at the transient snowline at the right moment one can make a good first guess of its position. The (transient) snowline marks the transition from last winter's snow to bare ice or to the snowpack of the previous winter. In the autumn and early winter the snowline can go downglacier in big steps (or in just one big step if a major snowfall covers the entire glacier at one time, which is not unusual). In spring the snowline moves upglacier more gradually as melt continues (*figure 2.3*). This will go on until the end of the ablation season. Because there will be no further melt, the snowline must now be close to the altitude where the winter accumulation is balanced by summer melt and thus marks the equilibrium-line altitude.

Figure 2.3. Picture of a small glacier in central Switzerland (Diablerets), taken in September 1984. The dark band running across the glacier is very dirty snow. It separates the bare ice surface from the snow surface and more or less marks the position of the equilibrium line.

Note, however, that this argument does not hold when a significant amount of melt water from the winter snowpack has frozen on the underlying ice to form a layer of superimposed ice. Superimposed ice formation is a common feature of subpolar glaciers, because the ice is cold, at least in early summer. However, on so-called temperate glaciers it is an insignificant process (by definition, temperate glaciers are at melting point almost everywhere).

The idea that the highest position of the snowline marks the equilibrium line has been used to map changes in E from air photography [4]. Clearly, the advantage of such a method is that a large area can be investigated rather quickly. However, the error bars are large. Early snowfalls, for instance, may confuse the picture considerably.

Are variations in E related to variations in the mean mass balance of a glacier? This point has been investigated for a number of glaciers on which extensive mass balance programmes are running. A study of interannual variability normally reveals a significant relation between E and B_m. For a simple glacier geometry (constant glacier width, constant slope) and a specific balance that increases linearly with height we should find

$$B_m = c\,E \; . \qquad (2.3)$$

Earlier studies have shown that eq. (2.3) holds for most glaciers [5]. *Figure 2.4* shows data for four glaciers.

Apparently, the slope of the regression line is more or less the same for Nigardsbreen, Hintereisferner and Abramov Glacier. The fact that it is different for White Glacier is related to the balance gradient (the rate at which the specific balance changes with altitude). On White Glacier this gradient is small, as is normal on dry polar glaciers and ice caps.

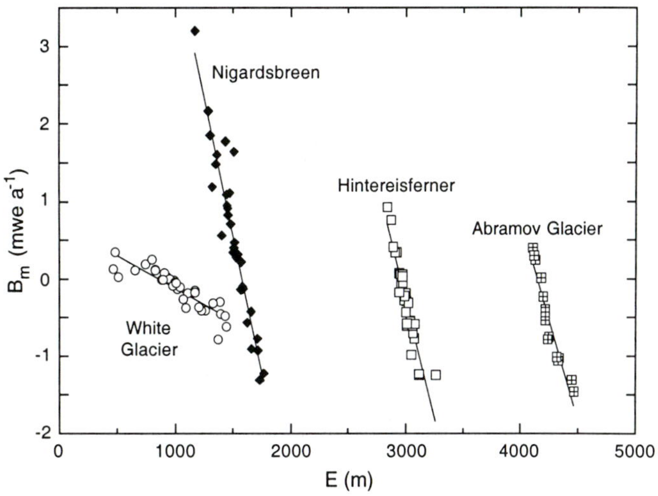

Figure 2.4. A plot of annual values of B_m versus E for four glaciers (White Glacier in Canadian Arctic, Nigardsbreen in Norway, Hintereisferner in Austria, Abramov Glacier in central Asia). Data from the World Glacier Monitoring Service [6] with some additions.

2.3 The thermal regime

Beacuse the temperature in the atmosphere decreases with altitude, snow settling on the higher parts of glaciers is generally colder than snow settling further down. The density of freshly fallen snow depends on temperature and wind conditions. A typical value for fresh snow on mid-latitude glaciers is 100 kg m^{-3}.

Then a complicated process starts in which the snow crystals change their shape due to mechanical action by the wind, water vapour transport in the snowpack and melting and refreezing. After many cycles of melting and refreezing the snow turns into firn. According to the official terminology, firn is snow that is at least one year old. However, skiers know better. In spring and early summer sunny slopes have good firn produced from snow that has fallen months or even only weeks earlier. The crystals are large and rounded and the density is typically 500 kg m^{-3}.

On mid-latitude and tropical glaciers, firn takes one to five years to change into ice, depending on the local micro-meteorological conditions. This is in sharp contrast to the hundreds of years needed to transform snow into ice on the Greenland and Antarctic ice sheets.

This book focuses on temperate glaciers. As mentioned earlier, these are glaciers that are at melting point almost everywhere. Most of the glaciers outside the polar regions are temperate. At first glance this is remarkable because the bulk of most glaciers is located far above the 0°C level of the annual atmospheric temperature. However, a glacier does not simply adjust to the mean air temperature; it is also subject to many thermodynamic processes that affect its thermal structure. In most cases refreezing of melt water is the dominant process. In spring, when the energy budget of the snow surface becomes positive and melt water forms, the temperature of the snowpack is normally well below freezing point. As a consequence, melt water penetrating into the snow refreezes and a large amount of heat is released. This process occurs everywhere below the dry snow line.

Further down the glacier more energy becomes available for melting and the snow is not as cold as higher up. The heating of the snowpack therefore proceeds much faster.

As an illustration, *figure 2.5* shows snow and air temperature data measured by an automatic weather station (AWS) on the tongue of the Morteratschgletscher (see Chapter 3) in the winter of 1995/96. As well as various other meteorological quantities, this station measures air temperature at a height of about 3 m. Snow depth is measured with an ultrasonic ranger. Two temperature sensors were inserted in the snow, 15 and 30 cm above the ice surface, after the first major snowfall. It can be seen that snow temperature follows the fluctuations in air temperature, but with a phase lag and smaller amplitude. Later in winter the snow temperature fluctuates less because the distance between snow surface and sensors is larger.

The first melt water occurs on March 23rd, but it is not enough to raise the temperature of the snowpack to melting point. This happens on April 10th .

When the snow pack has heated up and the ice underneath is not cold, runoff may start and the amount of snow diminishes rapidly. In the case shown the entire snow cover disappears within some weeks (in spite of the fact that 25 cm of fresh snow fell on May 3rd).

The importance of refreezing in the heat budget of a snowpack can be illustrated by a simple calculation. The latent heat of melt, denoted by L_m, is $3.34 \cdot 10^5$ J kg^{-1}. Because the specific heat is about 2100 J kg^{-1} K^{-1}, heating of 1 kg of snow (or ice) by 1 K requires the same amount of energy as melting 0.0063 kg of snow. A 2 m winter snowpack of uniform density at -10 °C can therefore be brought to melting point by the melting and refreezing of only 12.5 cm of the snowpack.

The structure of the upper layers of a glacier is intimately related to the thermal regime. The 'simple zones' are the dry snow zone, where no melt water is ever produced, and the lower ablation zone, where the winter snow melts away rapidly and ice melts over a long period of time. The data shown in *figure 2.5* are from the lower ablation zone. In between are the 'complex zones', as sketched in *figure 2.6*.

Most valley glaciers outside the Arctic and Antarctic regions are temperate. Temperatures below melting point are found only in very high places. Just below the equilibrium line temperatures a few degrees below freezing point may occur in a thin layer just below the surface. This layer hardly affects ice flow, because the cold ice does not extend to the glacier base.

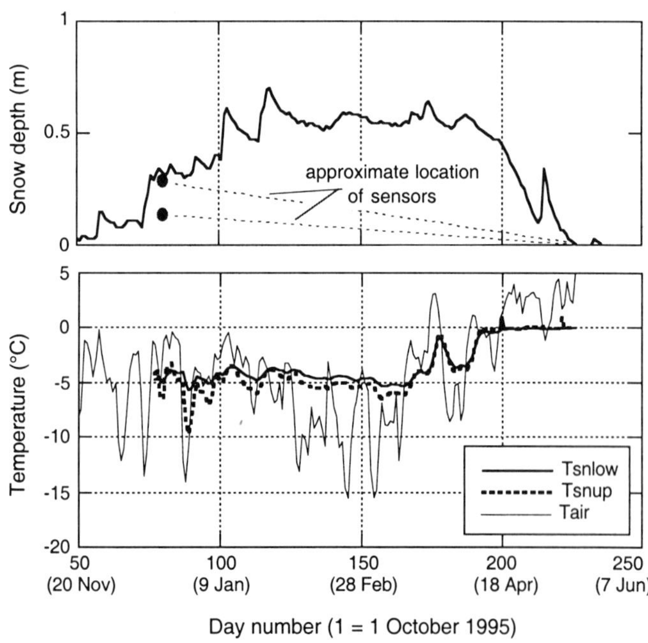

Figure 2.5. Snow temperature measured on the Morteratschgletscher, Switzerland, winter 1995/96. The upper panel shows snow depth measured with a sonic ranger. The approximate positions of the temperature sensors (lower: Tsnlow; upper: Tsnup) are indicated by the dashed lines.

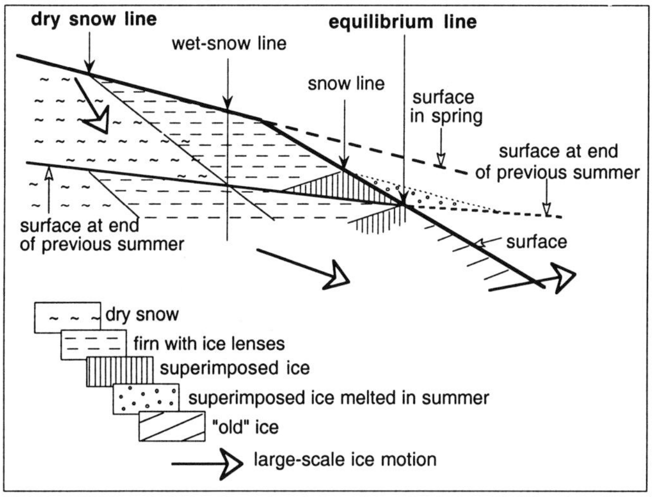

Figure 2.6. A diagram depicting the different zones on a glacier, in which the horizontal scale is typically a hundred times larger than the vertical scale. Conditions may vary strongly from year to year, and no doubt reality is even more complex. Large-scale ice motion keeps the surface in place.

2.4 Flow of ice

Anyone looking at a large valley glacier will realise that the ice flows downwards and that in one way or another the ice is able to deform plastically. Many features on the surface and at the sides express the basic law: gravity pulls the ice down, friction at the base and the sides attempts to keep the ice in place. One would expect deformation of the ice body to act in such a way that the internal stresses are minimised. Nevertheless, at many places this does not prevent stresses from becoming larger than the failure threshold and crevasses are formed. On the larger scale, however, deformation and crevasses allow the glacier to assume a fairly regular surface. The surface is a smoothed image of the underlying bed.

When studying the relation between glacier fluctuations and climate, one is interested primarily in the total mass budget of a glacier. But to what extent do glacier mechanics and mass budget interact? Do we need a sophisticated description of the stress and flow field in a glacier to predict its response to climate change? At this stage it is difficult to give a clear answer to this question, but a few simple observations are appropriate here.

Suppose that a glacier is in balance when a sudden drop in the equilibrium-line altitude occurs. The accumulation area will increase and the ablation area will decrease. The glacier will therefore gain mass and expand to make the ablation area larger again until a new balance is reached. This can all be handled with purely geometrical considerations. However, when the area of the glacier increases the ice thickness will increase as well. This is a dynamic process that can only be modelled by taking ice mechanics into account. Larger ice thickness implies a shift of the equilibrium line in the direction of the glacier terminus; this in turn increases the accumulation area and decreases the ablation area. Thus the dynamic response of a glacier provides feedback to the mass balance in a positive way. As we will see later, when the slope of the bed is small the feedback becomes very strong. It increases the sensitivity of a glacier to climate change and leads to larger response times. In conclusion we can say that in a glacier model for climate change studies the characteristic ice thickness should be able to respond to changes in glacier area and length.

It is general practice in the description of glacier flow to distinguish between two modes of movement. First of all there is internal deformation, involving the continuous displacement of ice crystals relative to each other. In larger glaciers the type of deformation approaches simple shearing flow, in which the ice velocity is more or less parallel to the bed and increases with height above the bed and with distance from the valley walls, see *figure 2.7*.

The deformation of (fictitious) parallel ice layers induces shear stresses at the interface of the layers. This shear stress must increase with depth, as it has to hold more and more ice in balance (the whole column of ice on which the body force, gravity, is working).

The second mode of movement is sliding: the glacier slides over the bed. However, because of the finite size and the ever present irregularities at the bed, this sliding must of course involve deformation as well. Otherwise the glacier could not maintain its shape. This deformation is of a three-dimensional nature. On temperate valley glaciers the contributions of sliding and deformation to the total mass flux are comparable.

A glacier may slide over bumps in the bedrock because of a special process called regelation. Ice melts due to increased pressure on the upstream side of the bump. The melt water flows around the obstacle and refreezes on the downstream side of the bump. The early theories of sliding dealt with this process. However, irregular short-term changes in ice velocity, observed on many glaciers, cannot be explained by regelation. Here other processes, like the formation of cavities and changes in the hydraulic system, play an important role.

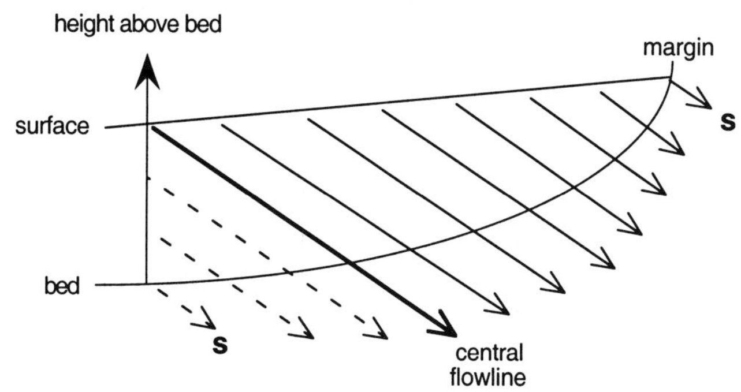

Figure 2.7. Motion in a valley glacier as depicted by velocity vectors in half a cross section. Solid arrows depict surface velocity, dashed errors show the variation with depth. Sliding (or slip) is indicated by s.

2.5 Mass balance and continuity

In the end, the glacier model we want to construct in the course of this text will be in a Eulerian framework. In general we will use a Cartesian coordinate system (x,y,z), in which x and y are the horizontal coordinates (map coordinates) and z is pointing upwards (along the local vector of gravity). The evolution of a glacier is then described in terms of a rate equation for ice thickness H:

$$\frac{\partial H(x,y)}{\partial t} = ... \qquad (2.4)$$

Here t denotes time. The right-hand side of eq. (2.4) should come from mass continuity in one way or another. Its final form will depend on the geometry of the model.
In general, the conservation of mass for a continuum can be formulated as

$$\frac{\partial \rho(x,y,z)}{\partial t} = -\nabla \bullet [\rho(x,y,z)\ \mathbf{v}(x,y,z)] \ . \qquad (2.5)$$

Here ρ is ice density and **v** the three-dimensional ice velocity vector. Now the simplest way of deriving the right-hand side of eq. (2.4) from this is to assume that ice density is constant. Integration over the vertical then yields (e.g. [8]):

$$\frac{\partial H}{\partial t} = -\nabla \bullet \mathbf{F} + B \ . \qquad (2.6)$$

F is the vertically integrated volume flux (a two-dimensional vector in the x,y-plane) and B the specific balance rate.
The principle of using the vertically integrated continuity equation to describe the evolution of a glacier is illustrated in *figure 2.8* for the case with one horizontal coordinate.

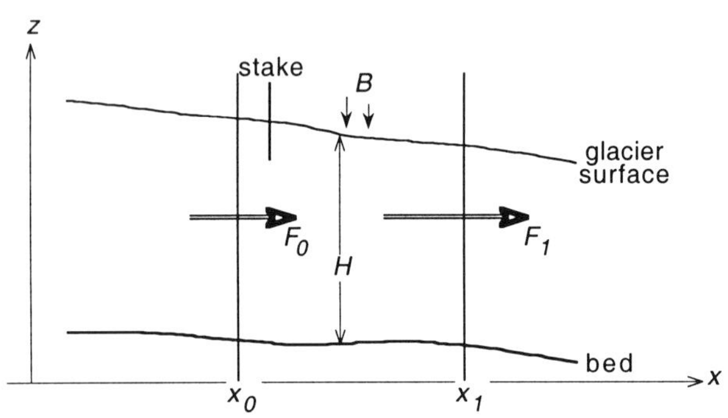

Figure 2.8. Continuity in a vertically integrated ice-flow model, illustrated in a finite-difference configuration. Per unit of time, the change in mean ice thickness between x_0 and x_1 should equal the difference in volume flux (F_1-F_0) plus the gain or loss at the surface (which is the specific balance B). In the case shown, divergence of mass by ice flow is (partly) compensated by gain at the surface (B>0). A stake is also shown. Readings for the exposed part yield an estimate of B, but do not give information about the change in ice thickness.

By introducing a vertical mean horizontal ice velocity **U** (also a two-dimensional vector in the x,y-plane), we can also write eq. (2.6) as:

$$\frac{\partial H}{\partial t} = -\nabla \bullet (H\,\mathbf{U}) + B \ . \qquad (2.7)$$

Eq. (2.7) can be used to calculate the evolution of a glacier when the balance distribution is known and when **U** can be related to the geometry of the glacier. This is not a trivial matter, of course. In a later chapter it will be argued that for larger valley glaciers **U** can be related to the local ice thickness and local surface slope. Then a substantial simplification of the description of ice mechanics becomes possible.

2.6 Traces in the landscape

Large parts of the earth's surface have been shaped by glaciers and ice sheets in one way or another. Glaciers have a tremendous erosive capacity, as is illustrated by the deep fjords found in all regions where during the last few millions of years glaciation has been the rule rather than the exception. But glaciers also deposit material in many ways. Terminal moraines, for instance, are the best indicators of former maximum stands of glaciers.
A number of excellent books have been written about glacial geomorphology [9] and interested readers will easily find their way to these. For the study of glacier fluctuations and climate change the most important elements are:

- trimlines and side moraines,
- terminal moraines.

Terminal moraines are formed by a glacier snout that is advancing and then starts to retreat after some time. An advancing glacier will destroy most of the terminal moraines that were deposited earlier. In contrast, a glacier in general retreat may leave behind a series of terminal moraines associated with short periods of advance. Some slowly retreating glaciers with a strong seasonal variation in ice velocity deposit a small terminal moraine every year.
Many glaciers in the Alps show a series of terminal moraines. They have been eroded partly by streams and are sometimes not recognised by the inexperienced observer. The patterns are different for different glaciers, of course. Nevertheless, there is a well-defined common element. Larger glaciers in the Alps show marked moraines deposited in the years ~1650, ~1850 and ~1920. For some glaciers the ~1650 terminal moraine was overridden by the ~1850 advance. The ~1920 advance is not so marked for all glaciers.
Side moraines are also interesting because they provide a rough indication of ice thickness and ice extent in earlier times. Some moraines are very impressive, like the complex shown in *figure 2.9*.

Figure 2.9. Huge side moraines formed by the Kviåarjökull, Iceland. This glacier flows down from Öræfa (2119 m), which is the highest mountain in Iceland. The moraines (there is a similar one on the left-hand side of the glacier) which extend onto a coastal plain, are about 3.5 km long and about 100 m high. They were formed during several high stands of the glacier in the Holocene. Photograph taken in June 1996.

When a glacier advances into a valley and becomes thicker and thicker it will scrape away all vegetation and surface irregularities from the valley walls. After the retreat of the glacier trimlines are left behind. These mark the upper limit of the erosive action and are very useful for estimating former ice levels. In fact, some glaciologists have argued that trimlines are not useful for this purpose because kinematic waves travelling downglacier could well disturb the picture. However, there is little evidence to support this view [10].

Altogether, the traces of maximum glacier stands are easily seen in the field (*figure 2.10*). Many glaciers in the world reached their neoglacial maximum only one or two centuries ago, and are now retreating. Therefore glacier behaviour provides very good source material to study the magnitude and extent of current global warming.

2.7 Historical glacier records

Historical glacier records will be used extensively in this book. They show that valley glaciers fluctuate considerably on a decadal-to-century time scale and are therefore a good basis for studying processes that relate glacier mass-balance to climate. Historical glacier records may be derived from different types of information.

Figure 2.10. The snout of Morteratschgletscher, Switzerland, and its forefield. This glacier is now about 7 km long. Since 1870, it has retreated over a distance of about 2 km. At the time of maximum extent the glacier front was located only a few hundred metres from the Landgasthof Morteratsch (the little white block on the right). Forest has started to grow on ground that was covered by the glacier at the beginning of this century. In the far distance the village of Pontresina can be seen. Photograph taken from Munt Pers (3207 m) in September 1997.

Figure 2.11. The oldest known drawing of a glacier (1601). The advancing Vernagtferner, Oeztal Alps, Austria, has blocked the valley and a huge lake has formed. Sooner or later the ice dam will break, causing a catastrophic flood. In the course of history this happened several times, bringing about devastation in the Oeztal [11].

Written documents going back to the 16th century tell about catastrophic floods caused by glacier-dammed lakes. In fact, the oldest known drawing of a glacier, shown in *figure 2.11*, was produced to explain how glacier floods come about [11].

Some documents contain reports on arable land and farms destroyed by advancing glaciers, for instance in the 18th century in Norway [12].

In addition, a wealth of information is available from sketches, etches and paintings of glaciers. There are a few thousand of these, but many show the same glaciers. The number of glaciers for which two or more useful pictures from different times in the 18th or 19th century exist is small (maybe a few dozen).

Systematic mapping of glaciers started only 100 years ago and has been limited to a few glaciers. The most comprehensive data set is related to length variations. It is relatively easy to make annual observations regarding the length of a valley glacier. Volunteers from Alpine clubs and other organisations have played an important role in compiling an impressive data set.

From the above it will be clear that data on historical glacier fluctuations are mainly restricted to glacier length, and that even these records are inhomogeneous. The record of Nigardsbreen is a typical example (*figure 2.12*).

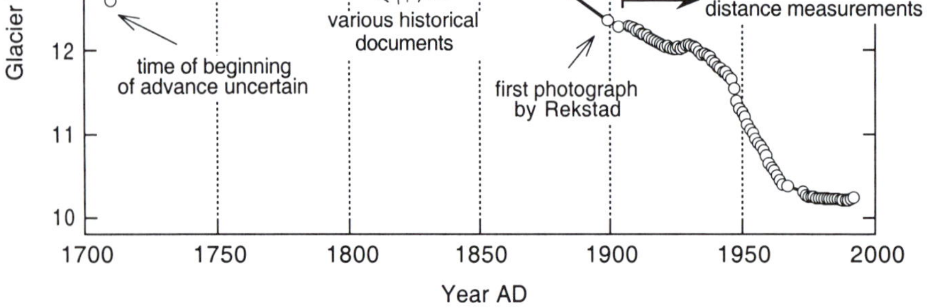

Figure 2.12. Example of a historical glacier record: the length record of Nigardsbreen, Norway. Different types of information are combined to make one record [13].

References and further reading

1. Z. Zuo and J. Oerlemans (1997): Contribution of glacier melt to sea-level rise since AD 1865: a regionally differentiated calculation. *Climate Dynamics* **13**, 835-845.

2. W. Haeberli, H. Bösch, K. Scherler, G. Østrem and C.C. Wallén (1989) *World Glacier Inventory. Status 1988.* IAHS(ICSI)-UEP-UNESCO.

3. M.B. Dyurgerov and M.F. Meier (1997): Mass balance of mountain and subpolar glaciers: a new global assessment for 1961-1990. *Arctic and Alpine Research* **29**, 379-391.

4. T.H.H. Chinn (1989): Glaciers of New Zealand. In: *Satellite Image Atlas of Glaciers in the World; Irian Jaya, Indonesia, and New Zealand.* U.S. Geological Survey Professional Paper 1386-H.

5. R.J. Braithwaite (1984): Can the mass balance of a glacier be estimated from its equilibrium-line altitude? *Journal of Glaciology* **30,** 364-368.

6. P. Kasser (1967) *Fluctuations of Glaciers, 1959-1965.* (Vol. I) IASH (ICSI)- UNESCO.

   P. Kasser (1973) *Fluctuations of Glaciers, 1965-1970.* (Vol. II) IASH (ICSI)- UNESCO.

   F. Müller (1977) Fluctuations of Glaciers,1970-1975. (Vol. III) IASH (ICSI)- UNESCO.

   W. Haeberli (1985): *Fluctuations of Glaciers, 1975-1980.* (Vol. IV) IASH (ICSI)- UNESCO.

   W. Haeberli and P. Müller (1988): *Fluctuations of Glaciers, 1980-1985.* (Vol. V) IASH (ICSI)- UNEP-UNESCO.

   W. Haeberli and M. Hoelzle (1993): *Fluctuations of Glaciers, 1985-1990.* (Vol. VI) IASH (ICSI)- UNEP-UNESCO.

   W. Haeberli, M. Hoelzle, S. Suter and R. Frauenfelder (1998): *Fluctuations of Glaciers, 1990-1995.* (Vol. VII) IASH (ICSI)-UNEP- UNESCO.

7. C.S. Benson (1962): *Stratigraphic studies in the snow and firn on the Greenland ice sheet.* SIPRE Research Rep. 70, U.S. Army Corps of Engineers, Wilmette, Illinois, 93 pp.

8. J. Oerlemans and C.J. van der Veen (1984): *Ice Sheets and Climate.* Reidel, 217 pp.

9. D.E. Sugden and B.S. John (1976): *Glaciers and Landscape. A Geomorphological Approach.* Edward Arnold (London), 376 pp.

   D. Drewry (1986): *Glacial Geologic Processes.* Edward Arnold (London), 276 pp.

10. R.S.W. van de Wal and J. Oerlemans (1995): Response of valley glaciers to climate change and kinematic waves: a study with a numerical ice-flow model. *Journal of Glaciology* **9**, 115-131.

11. K. Nicolussie (1990): Bilddokumente zur Geschichte des Vernagtferners im 17. Jahrhundert. *Zeitschrift für Gletscherkunde und Glazialgeologie* **26** (2), 97-119.

12. G. Østrem G, O. Liestøl and B. Wold (1977): Glaciological investigations at Nigardsbreen, Norway. *Norsk GeografiskTidsskrift* **30**, 187-209.

13. J. Oerlemans (1997): A flow-line model for Nigardsbreen: projection of future glacier length based on dynamic calibration with the historic record. *Annals of Glaciology* **24**, 382-389.

3. The micro-climate of a valley glacier

3.1 Introduction

The surface of a glacier differs from its surroundings. The contrast may be small in winter when the landscape is totally covered by snow. However, in summer the differences can become very large. Those parts of a glacier that are not covered by debris have a relatively high albedo. As a consequence, less solar energy is absorbed. The temperature of the glacier surface cannot rise above melting point, so on warm days it is a sink of energy for the overlying air.

On warm summer days alpine valleys heat up considerably. Slope winds help to distribute the warm air generated at the valley walls that face the sun. This process is quite effective and the air in a valley is rapidly mixed (characteristic time scale: 15 minutes). Due to differential heating along the valley axis, an up-valley flow is normally established: the valley wind [1]. The layer of air involved is typically 1 km thick and blows up-valley. The valley wind brings heat to melting glaciers. One would expect valley winds to become weaker when they reach the glacierised part of a valley. In fact, the situation may become quite complex because glacier ice and walls of bare rock may be distributed rather irregularly.

At night the air is cooled and a mountain wind flows down the valley. It is much weaker and shallower than the valley wind, because turbulent exchange between surface and air is suppressed by the stable stratification.

Systematic investigations of the heat budget of a melting glacier did not start until after the second world war. Pioneering work was done by Ambach [2]. A series of other investigations followed [3], and these studies have given us basic insight into the nature of the energy budget on glaciers.

In more recent years glacio-meteorological experiments have been done in which the components of the surface energy flux were measured simultaneously at a large number of stations [4,5,6]. Large data sets have been obtained, providing insight into altitudinal gradients of meteorological quantities. Because a great deal of the data used in this chapter were collected during these experiments, an overview is given in *table 3.1*.

The melt zones studied in these summer experiments differ in size and in their climatic setting. With regard to the transect on the Greenland ice sheet there is lateral symmetry to a large extent. The climatic conditions are dry (annual precipitation of the order of 0.3 m), relatively sunny and with a large annual temperature range (of about 28 K). The melt season is short.

The situation on the Pasterze, a 10 km long valley glacier in the Austrian Alps, is very different: the flow is channelled by the valley walls. The melt season is much longer and the annual amount of melt is typically 5 m.

The climatic conditions on Vatnajökull, which is the largest ice cap in Iceland (about 8000 km^2), are very maritime: high precipitation, high melt rates and a relatively small seasonal cycle in temperature.

During PASTEX-94, about 200 ascents were made with a cable balloon from the glacier. The sensor mounted on the balloon measures wind speed, wind direction, pressure, humidity and temperature. Graphs from a single ascent made on a day with fair weather illustrates the type of information that can be obtained (*figure 3.1*, page 17). The most outstanding feature is the sharp temperature increase just above the glacier surface, about 14 K within 20 m. The glacier wind has a maximum speed of about 5 m s^{-1}; the height of the wind maximum is just a few metres above the surface. At a height of about 80 m wind speed is low and the wind direction turns from down-glacier (~340°) to up-glacier (~160°). Higher up the valley wind blows at a typical speed of 2 m s^{-1}. In the case shown the large-scale flow was very weak.

location	period	equipment	participants
GIMEX-91			
Central West Greenland	summer 1991	7 weather stations	IMAU, Utrecht University
0 - 1000 m		RASS-SODAR	Free University Amsterdam
L ~ 100 km		cable balloon	
		eddy correlation meas.	
		31 m profile mast	
PASTEX-94			
Pasterze, Austria	summer 1994	7 weather stations	IMAU, Utrecht University
2000 - 3400 m		cable balloon	Free University Amsterdam
L ~ 10 km		eddy correlation meas.	
		13 m profile mast	
VATEX-96			
Vatnajökull, Iceland	summer 1996	12 weather stations	IMAU, Utrecht University
0 - 1600 m		radiosonde	Free University Amsterdam
L ~ 50 km		cable balloon	University of Iceland
		eddy correlation meas.	University of Innsbruck

Table 3.1. Recent glacio-meteorological experiments in which the micro-climate of melting glacier surfaces has been studied. Acronyms on the left are used in the text. At all weather stations profile measurements were made of meteorological quantities. The altitudinal range and the horizontal length scale covered by the weather stations are given below the location.

The observations reveal that in summer a circulation system, with a glacier wind driven by cooling of the air just above the glacier surface and a valley or synoptic-scale wind above, is present most of the time. This system is not restricted to valley glaciers but has also been observed over the melt zones of Vatnajökull (Iceland) and in West Greenland [4, 6]. The glacier wind may be an important mechanism for supplying heat to the melting glacier surface. Its structure and dynamics will therefore be discussed in some detail later.

The deposition of snow on a glacier is directly related to the synoptic-scale flow and its interaction with the orography. Ice caps and ice sheets create part of this orography, but this is hardly the case for valley glaciers. The growing or shrinking of valley glaciers does not interact directly with the precipitation pattern, although one can argue that smooth glacier surfaces facilitate the occurrence of drifting snow.

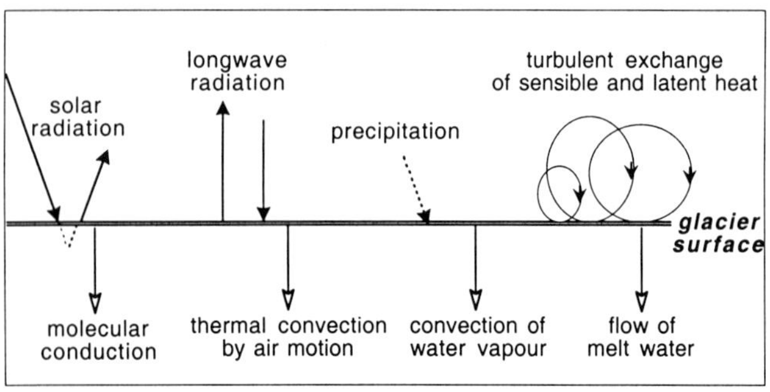

Figure 3.2. The most important processes determining the energy flux at the glacier - atmosphere interface and the thermal structure in the upper layer of the glacier.

For the melt process the surface energy budget is decisive. The mechanisms of heat transfer thought to be important for glaciers are shown in *figure 3.2.* The largest fluxes are the radiative fluxes, typically a few hundred Watts per square metre. A large part of the solar radiation reaching the glacier surface is reflected - more in the case of fresh snow, less in the case of old snow or bare ice, very little when the surface is covered by morainic material. Solar radiation penetrates into the snow and ice. In fact, the reflection as measured at the surface is the result of a complicated scattering process at the ice or snow crystals in the upper metres of the glacier. For long wave radiation the surface is very dense. It has an emissivity (and absorptivity) of about one. The amount of longwave radiation reflected by the surface is negligible - it is all absorbed. Incoming and outgoing longwave radiation compensate each other to a large degree. Normally the amount emitted by the surface is slightly larger (a few tens of Watts per metre squared) than the amount coming in from the atmosphere, making the net longwave balance negative. However, when the air is warm and humid, and clouds are present, the longwave balance may be positive.

The effect of clouds on the shortwave and longwave radiation budgets is of opposite sign. More clouds implies less shortwave radiation and more longwave radiation. The net effect depends on the surface albedo and on the cloud transmissivity [7]. For a high surface albedo (e.g. fresh snow) the net change in longwave radiation for a given increase in cloudiness is larger than the net change in the solar radiation. Consequenty, the net radiation balance increases. For lower albedo (e.g. ice) the solar radiation effect dominates and the net radiation budget decreases with cloudiness.

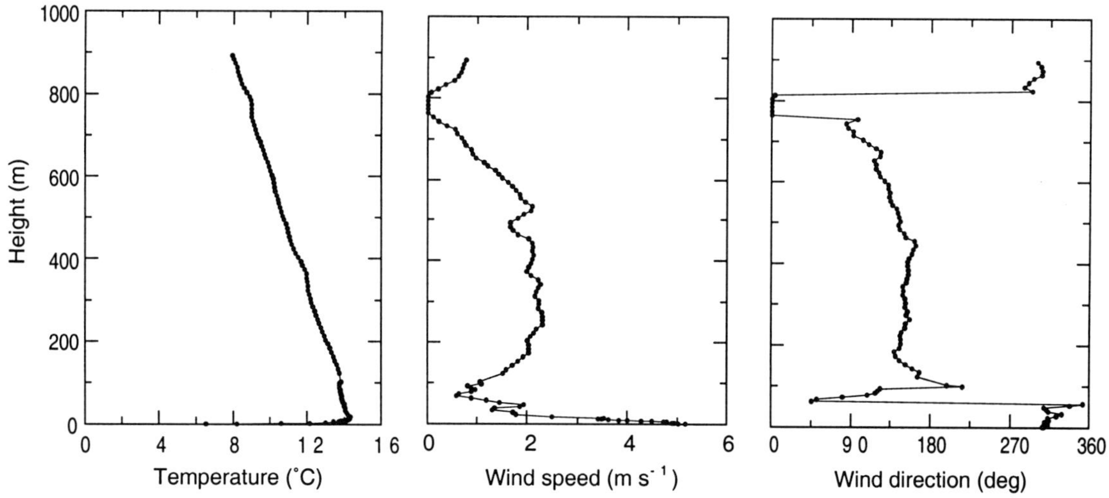

Figure 3.1. Structure of the atmosphere above the melting glacier tongue of Pasterze, Austria, as measured with a balloon. The ascent was made in fair weather on 23 July 1994, 16:25 local time. Note that the surface temperature is 0°C, implying a very large temperature gradient in the lower metres.

Turbulent exchange of heat and moisture can be quite significant, especially in winter (when the sun is low) or in summer when air temperature is high, as in the case shown in *figure 3.1*. The fluxes are in the direction of the gradients in the mean temperature and humidity profiles. When air temperature is above freezing point the sensible heat flux is always towards the surface. In such conditions the latent heat flux can go in both directions, depending on the humidity of the air. The saturation vapour pressure e_s for a melting glacier surface is 610.8 Pa. If an air mass has a temperature of 10°C, the gradient in vapour pressure and hence the vapour flux changes sign at a relative humidity of about 50%. At lower humidity evaporation cools the surface, at higher humidity condensation heats the surface.

The temperature of the precipitation may differ from the temperature at the surface. This implies that precipitation may add or remove heat from the existing snowpack or ice surface. However, the energy fluxes involved are small.

The fluxes inside the glacier are much smaller than those between atmosphere and glacier surface, except for the flow of melt water, which represents a latent heat flux. When glacier ice is at the surface molecular conduction is the only important process (note, however, that in the interior of cold glaciers conduction, frictional heating and advection due to ice motion have a similar order of magnitude).

In snow and firn, convection by air motion transports heat and water vapour. The fluxes are small but important for the metamorphosis of snow crystals. The vapour transport is determined mainly by the density of the snow and the temperature gradient. Fairly comprehensive models have been developed to describe the metamorphosis of snow in different climatic conditions; most of these models are connected with avalanche research.

For the study of the mass balance of temperate valley glaciers a description of the surface energy flux Ψ (in: W m^{-2}) is required. We write:

$$\Psi = Q\,(1-\alpha) + L_{in} + L_{out} + H_S + H_L + G\,. \tag{3.1}$$

Fluxes towards the glacier surface are defined as positive. L_{in} and L_{out} are incoming and outgoing longwave radiation, respectively. H_S is the turbulent sensible heat flux, H_L is the turbulent latent heat flux. G is the heat flux into or out of the glacier due to conduction and convection, and due to penetration of solar radiation. The first term on the right-hand side is the absorption of solar radiation. Q is the global radiation and α the albedo. Global radiation is defined as the amount of solar energy impinging on a horizontal surface coming in through a half-sphere. This implies that eq. (3.1) refers to a horizontal surface. The glacier surface is never horizontal and eq. (3.1) has to be modified if it is to be used for a real glacier.

Since radiation is a dominating process considerable attention will be given to it in the following sections. It is useful to list a few basic equations for reference. First of all there is the Planck function, giving the spectral distribution of radiation emitted by a black body. Here it is expressed as the monochromatic radiance B_λ:

$$B_\lambda(T) = \frac{2\,h\,c^2}{\lambda^5\left[\exp(ch/k\lambda T) - 1\right]}\,, \tag{3.2}$$

where T is the temperature of the radiating body, and λ is the wavelength. The physical constants on the right-hand side of eq. (3.2) are:

$h = 6.63 \cdot 10^{-34}$ J s (Planck's constant),
$k = 1.38 \cdot 10^{-23}$ J K^{-1} (Boltzmann's constant),
$c = 2.9973 \cdot 10^{8}$ m s^{-1} (speed of light in vacuum).

From eq. (3.2) Wien's law can be derived:

$$T\,\lambda_{max} = 2898\ \mu m\ K\ . \qquad (3.3)$$

This expression shows that there is a simple relation between the radiative temperature and the wavelength of maximum emission: higher temperature implies a smaller wavelength. As characteristic values of the radiative temperature of the earth and the sun we can use 293 and 6110 K, respectively. Substitution yields $\lambda_{max} = 9.9\ \mu m$ (earth) and $\lambda_{max} = 0.474\ \mu m$ (sun). By drawing and comparing the black body curves for solar and terrestrial radiation one can easily verify that the overlap is quite small. This justifies the distinction between shortwave and longwave radiation in the context of the surface energy balance.
Finally, the total amount of radiation emitted by a black body is obtained by integrating eq. (3.2) over wavelength and the hemisphere, yielding

$$I_{emitted} = \sigma T^4\ , \qquad (3.4)$$

where σ is the Stefan-Boltzmann constant (equal to 5.67×10^{-8} $J\ m^{-2}\ K^{-4}\ s^{-1}$).
A wealth of information concerning the role of radiation in the global climate system, including a discussion about concepts and definitions, can be found in [8].

3.2 Solar radiation - geometric factors

The real spectrum of solar radiation as measured outside the earth's atmosphere differs from the Planck function. During its path through the atmosphere the spectrum is further modified by scattering and absorption by air molecules, aerosols and clouds. The relevant quantity with respect to the surface energy flux is the spectrally integrated solar radiation. The term albedo as used earlier is defined as the intensity of the spectrally integrated reflected solar radiation (upwards through a horizontal plane) divided by the intensity of the spectrally integrated global radiation (downwards through a horizontal plane). It is possible to characterise the surface by looking at the reflectance in distinct spectral bands of satellite sensors (e.g. as provided by the Landsat Thematic Mapper). We will come back to this in Chapter 12. But for the moment we are interested principally in the total amount of solar energy available for absorption by the surface.
The solar constant S is defined as the energy flux through a unit area perpendicular to the solar beam, at the mean earth-sun distance. Its value is $1368\pm3\ W\ m^{-2}$. At present, the earth-sun distance varies little over the year, because the eccentricity of the earth orbit around the sun (currently 0.0167) is small.
We now define the extra-terrestrial irradiance on a horizontal surface Q_E as

$$Q_E = S\left(\frac{\bar{r}}{r}\right)^2 \sin e = S^* \sin e\ . \qquad (3.5)$$

Here e is the solar elevation, r the actual sun-earth distance and $\bar{r}$ its annual mean value. The solar elevation is given by

$$\sin e = \sin\phi\ \sin\delta + \cos\phi\ \cos\delta\ \cos h\ , \qquad (3.6)$$

where ϕ is latitude, δ the solar declination and h the hour angle.
The calculation of solar radiation for a given latitude at a given time of day is a standard procedure which will not be explained here [9].
It is instructive to consider the effect of latitude on the extra-terrestrial irradiance, because it is one of the basic factors determining the length and intensity of the ablation season. Increasing latitude implies a smaller annual mean value, a larger seasonal amplitude and a smaller daily amplitude of Q_E. *Figure 3.3* shows how the daily mean value of Q_E, denoted by $Q_{E,day}$, varies through the year. In summer, values of $Q_{E,day}$ change little with latitude. Furthermore the well-known features are seen: the double maximum in the equatorial regions and the polar night at high latitudes.
It should be noted that there is no simple relation between Q_E and global radiation. Because the mean solar elevation is less at higher latitudes, the apparent optical mass of the atmosphere is larger and absorption and reflection of solar radiation are stronger.
The seasonal cycle increases with latitude but the daily cycle decreases (*figure 3.4*). Nevertheless, it is only at very high latitudes that the daily cycle in Q_E becomes really small.
A number of steps are needed to relate Q_E to the solar radiation that may become available at a glacier surface. First of all there are local geometric factors like shading by mountains, and the slope and aspect of the surface and reflection from surrounding slopes.

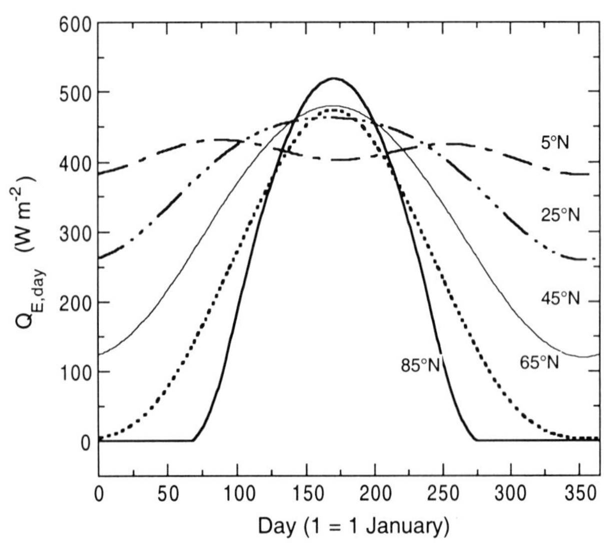

Figure 3.3. Daily mean values of the extra-terrestrial irradiance for a horizontal surface.

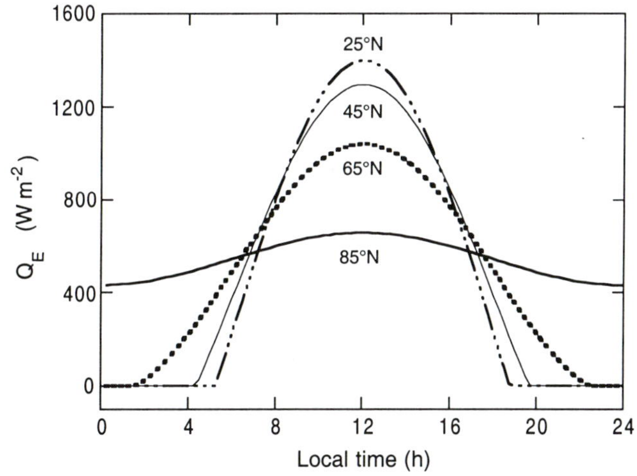

Figure 3.4. Extra-terrestrial irradiance for calendar day 181 at some selected latitudes.

To calculate the effect of geometry, the modification of the solar beam in the atmosphere has to be considered. Sophisticated methods have been devised to deal with this; the ultimate method for a valley with a complicated geometry in which scattering clouds are present is a Monte-Carlo technique which traces the path of millions of photons.
However, here we discuss the simplest approach. The global radiation is split into a direct part Q_{dir} and diffusive part Q_{dif} which is isotropic:

$$Q = Q_{dir} + Q_{dif} = f_{dir}\,Q + f_{dif}\,Q\,, \qquad (f_{dir} + f_{dif} = 1)\,. \tag{3.7}$$

For clear-sky conditions f_{dif} is typically 0.15, for an overcast sky 0.85 (although this really depends on the type of clouds: e.g. for a uniform layer of cirrostratus f_{dif} can be 0.5). The values of f_{dir} and f_{dif} also depend on other atmospheric constituents (notably aerosol), but clouds are the most important factor. A possible formulation is [10]:

$$f_{dir} = 0.2 + 0.65\,(1-n)\,, \qquad f_{dif} = 0.8 - 0.65\,(1-n)\,. \tag{3.8}$$

Here the fractional cloud cover is denoted by n.
The next step is to consider the amount of solar radiation impinging on a sloping surface. We denote the flux by Q_S, to distinguish it from Q, the flux through a horizontal surface. In our approach slope and aspect affect only the direct radiation because the diffusive part is assumed to be isotropic. Denoting the apparent solar elevation with respect to the sloping surface by e' we then have:

$$\begin{aligned} &\text{if } e > 0 \wedge e' > 0: \; Q_S = S^* (f_{dir} \sin e' + f_{dif} \sin e)\,, \\ &\text{if } e \geq 0 \wedge e' \leq 0: \; Q_S = S^* f_{dif} \sin e\,, \\ &\text{if } e \leq 0: \; Q_S = 0\,. \end{aligned} \tag{3.9}$$

This is about the simplest way in which the flux of solar radiation received by a sloping surface can be calculated from the local time, solar declination, latitude, cloudiness, and slope and aspect.

Now we turn to shading by surrounding mountains. The shading angle ϕ is determined by the skyline and of course varies with the azimuth. For many glaciers shading is a significant factor, or, put the other way round, glaciers tend to form in shaded places. Glaciers flowing into narrow deep valleys away from the sun can penetrate to much lower altitudes. An example is the Untere Grindelwaldgletscher, Switzerland (*figure 3.5*).
How can shading be dealt with? The approach in which the solar radiation is split into a direct and diffusive part is again useful. When the sun is behind the mountain rim the direct flux must vanish, but there will still be diffusive radiation as long as $e > 0$. So shading can be taken into account by setting the direct radiation to zero when $\phi > e$.

Figure 3.5. Untere Grindelwaldgletscher, Switzerland, flowing in a northwesterly direction. The snout of the glacier receives little solar radiation due to shading by the valley walls. In addition, the thick debris cover acts as an insulator and reduces the rate of melting. Photograph made in September 1991.

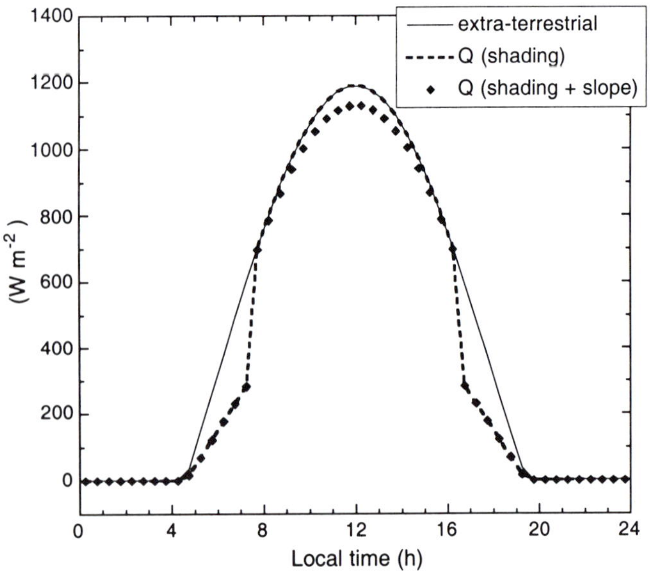

Figure 3.6. The effects of slope and shading for a cloudless day at 45°N according to eq. (3.9). The shading angle is 25°. The surface slopes 10° to the north.

Figure 3.6 shows an example for day 202 at a latitude of 45°N. The attenuation of the solar radiation in the atmosphere was not taken into account. This is somewhat inconsistent because it implies the presence of diffusive radiation in a fully transparent atmosphere. However, it allows us to study purely geometric effects. In the example shown the shading angle is 25° and the slope 10° to the north. Clearly, a considerable part of the extra-terrestrial irradiance does not reach the surface.

The effects of slope and shading are larger when the sun is low. Therefore at higher latitudes and in the transitional seasons the reduction of the solar flux at the surface will be larger than shown in *figure 3.6*.

We have now obtained a feeling for the importance of geometry. As mentioned earlier, this is only part of the game. Scattering and absorption of solar radiation in the atmosphere, notably by clouds, now need to be considered as well. In the following discussion we will use transmissivities to describe the attenuation of the solar flux in the atmosphere:

$$Q = \tau_e Q_E = \tau_{cl} \tau_a \tau_g Q_E \quad . \qquad (3.10)$$

Here τ_e is the total effective transmissivity, τ_{cl} the cloud transmissivity, τ_a the transmissivity related to absorption and scattering by air molecules, water vapour and aerosol. The geometric effects (shading, reflections from surrounding slopes) are represented by τ_g. Eq. (3.10) looks simple because the problems are hidden: the transmissivities are not independent. Geometric effects as discussed above depend on the cloud conditions and it is difficult to disentangle all the possible interactions. Nevertheless, to arrive at a workable approach that can be used in mass-balance models, eq. (3.10) is a useful starting point.

3.3 Solar radiation - atmospheric effects

The theory of scattering and absorption of solar radiation in the atmosphere is well developed [11]. Many algorithms have been constructed to calculate solar fluxes for a prescribed composition of the atmosphere. In general, the accuracy of calculations performed with such algorithms is limited by the quality of the input data, not by deficiencies in the radiative transfer theory.

Clouds form a particularly difficult aspect. In recent years, the application of satellite data has led to substantial progress in the quantification of cloud distribution and cloud physical properties. However, glacio-meteorology has not benefitted much. The detection of clouds over highly reflective surface from satellite data is much more difficult. For ice sheets and ice caps some information can be obtained, but there is a lack of data about valley glaciers. Weather satellites are not really helpful because the scale is too small.

From these considerations it will be clear that ground measurements of the solar flux on glaciers are very useful. For shorter periods of time these are available from the studies mentioned earlier. However, longer time series are also needed; these can only be obtained from automatic weather stations. Automatic weather stations are indeed used on glaciers and ice sheets. Most of them are in the accumulation zones, very few in the ablation zones. The reason is clear: ablation zones are difficult to work in because access is difficult and melt rates are high. Stations should be able to stay in place even when 5 to 10 m of ice melts away in one year.

Figure 3.7. The automatic weather station on the Morteratschgletscher, Switzerland, after an early snowfall in September 1998. The author is inspecting the datalogger and replacing the memory card. The station stands freely on the ice and has been running since October 1, 1995. On the right is an acoustic ranger mounted on stakes drilled into the ice. It measures the distance to the surface, from which melt and snow depth can be obtained.

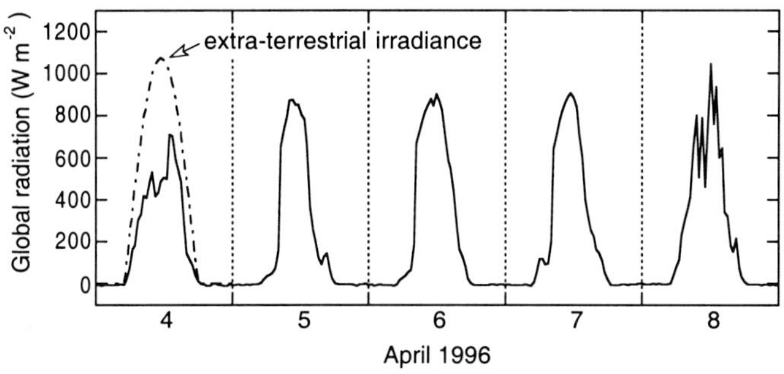

Figure 3.8. A record of global radiation from the tongue of the Morteratschgletscher, Switzerland [12].

In 1987 the Institute for Marine and Atmospheric Research, Utrecht University (IMAU), carried out the first tests with a construction which stood freely on the ice and which could therefore sink with the melting surface. The construction consists of a mast, free from the surface, to which four legs are attached (*figure 3.7*). These legs make a small angle with the surface (about 20°) and have pins at the end pointing downwards. After ten years of experience one can say that this construction works well. It has been used successfully in the meteorological experiments mentioned earlier and also for automatic weather stations that are designed to provide data all year round. The IMAU is now operating automatic weather stations in the melting zones of the Greenland Ice Sheet (western part), of Hardangerjökulen (Norway), and of the Morteratschgletscher (Switzerland). So far, the station on the Morteratschgletscher has delivered the best data set because it can be visited regularly and because atmospheric conditions are favourable (little icing).

Figure 3.8 shows a record of global radiation Q_d from the station at the Morteratschgletscher [12]. The station is located on the lowest part of the glacier, about 500 m from the glacier front, at an altitude of 2100 m. The surrounding mountains are typically 3000 - 4000 m high. The highest mountains are to the south and to the west, providing efficient shading with shading angles of up to 40°. In the example shown the 4th of April was a cloudy day, the period 5-7 had few clouds and on the 8th conditions were variable. On this day multiple reflections between clouds, snow-covered valley walls and glacier surface led to a peak value (half-hourly mean) that is comparable to the extra-terrestrial irradiance.

Figure 3.9 shows a 1-year record of daily mean global radiation (Q_d) from the Morteratsch station. Due to the high degree of shading the annual mean global radiation is only 49% of the extra-terrestrial irradiance. There is a strong day-to-day variation in Q_d, which is due almost entirely to varying cloud conditions. For the summer half year it is possible to draw an envelope that more or less marks the clear-sky global radiation (= $\tau_a \tau_g Q_E$, dashed line in the figure). For the winter and early spring this is not possible because in those periods there is a strong interaction between clouds and the snow-covered valley walls.

Averaged over the period 1 April - 30 September 1996, the mean value of Q is 203 W m^{-2}, and the estimated clear-sky global radiation is 290 W m^{-2}. In this case therefore clouds have reduced the global radiation by 30%. However, cloud observations are not available. To establish a more explicit relation between Q and cloudiness we have to look at more detailed measurements.

Greuell et al. analysed the PASTEX-94 data and considered the radiation balance in detail [13]. On the basis of 3-hourly cloud observations (208 data points in the period June 18 to August 9, 1994) the following relation was found for the cloud transmissivity at an altitude of 2205 m:

$$\tau_{cl} = 1 - 0.233\, n - 0.415\, n^2 \,. \qquad (3.11)$$

This relation is in very good agreement with an analysis of global radiation data from Austrian climate stations carried out by Sauberer [14]. His tables can be fitted well with (h is altitude):

$$\tau_{cl} = 1 - (0.41 - 6.5 \cdot 10^{-5}\, h)\, n - 0.37\, n^2 \,. \qquad (3.12)$$

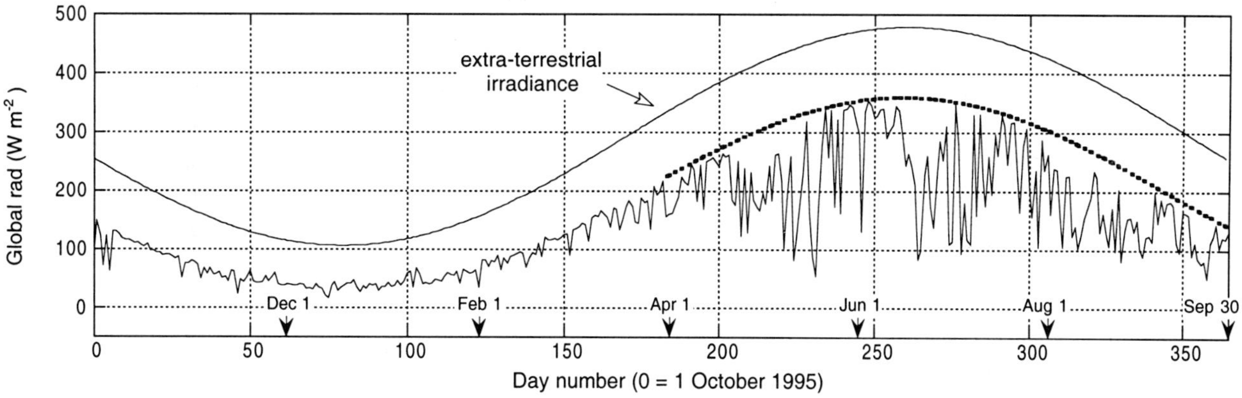

Figure 3.9. A one-year record of daily mean global radiation from the Morteratsch station. The data from the weather station have been corrected for systematic errors by comparison with calibrated instruments operated in parallel for shorter periods [12]. The dashed line is an attempt to estimate the clear-sky global radiation.

For the Greenland data Konzelmann et al. [15] find:

$$\tau_{cl} = 1 - 0.78\, n^2 \exp(-0.00085\, h)\ . \qquad (3.13)$$

Eq. (3.11) and similar fits for data from GIMEX-91 and VATEX-96 are shown in *figure 3.10*. From this picture it is quite clear that clouds on the Greenland ice sheet are optically thinner, especially at higher elevation. There is little effect of such clouds up to a fractional cloud cover of 0.3.

Altogether, the relations shown in *figure 3.10* give a good impression of how global radiation depends on cloudiness. The analysis is still primitive, because no distinction is made between different cloud types. There are many practical reasons why a more refined analysis is not feasible, at least not for valley glaciers. For larger ice masses like the Greenland and Antarctic ice sheets more detailed analysis is possible because there is generally more horizontal homogeneity in atmospheric conditions.

It is clouds which cause most of the temporal variability of global radiation, but other processes in the atmosphere like scattering and absorption by air molecules and aerosol contribute significantly to the overall depletion of solar radiation. The various constituents all have different scattering functions and absorption spectra, but a treatment of these goes beyond the scope of this book [11]. Instead we take a further look at data.

A survey of how global radiation in summer increases with altitude is presented in *figure 3.11*. Mean station data for GIMEX-91, PASTEX-94 and VATEX-96 are shown in one plot.

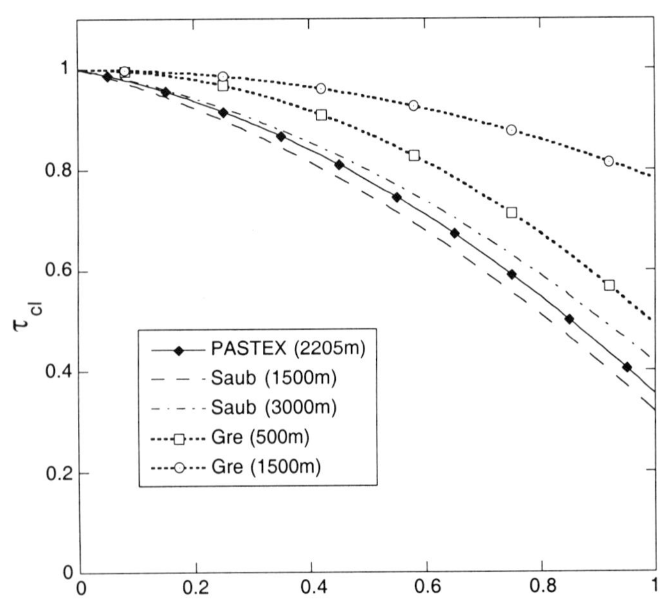

Figure 3.10. Parameterisation of cloud transmissivity based on the PASTEX-94 data. For comparison, fits of the tables of Sauberer are also shown for two altitudes [14]. The parameterisation for Greenland is also shown [15].

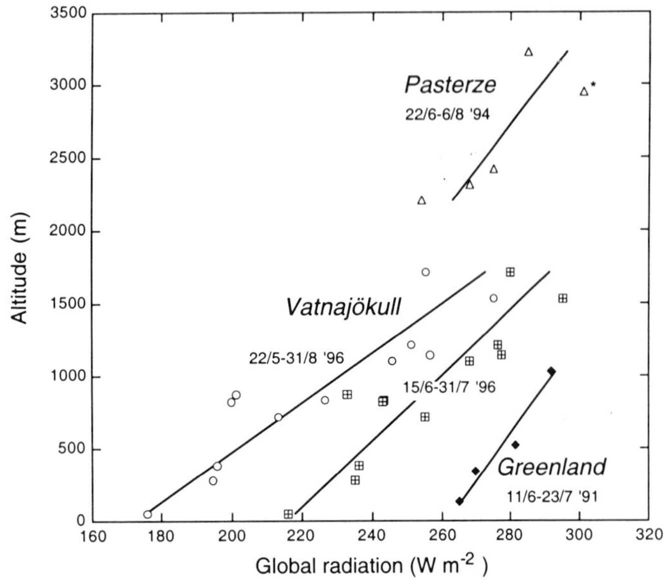

Figure 3.11. The increase in global radiation with altitude as measured during GIMEX-91, PASTEX-94 and VATEX-96. Mean values are shown over the periods indicated. For Vatnajökull two periods are shown, to facilitate a comparison with the data for the other glaciers. The Pasterze data point marked * is an outlier due to intense reflection from surrounding slopes [13]. Note that for Vatnajökull data are shown for a short and a long period in the summer of 1996.

First of all it is interesting to note that the values for the mean global radiation measured in high summer on the Pasterze and in Greenland are similar (260 - 300 W m^{-2}). Over the entire period of measurements (22 May - 31 August) the values from Vatnajökull are much smaller. When the period 15 June - 31 July is considered, a better comparison with the other data can be made. Nevertheless, the amount of global radiation in the lower reaches of Vatnajökull is relatively small.

A more detailed analysis of the data reveals that the differences seen are again due mainly to differences in cloud amount and cloud type. This applies to the characteristic values of global radiation as well as to their altitudinal gradients. In general, clouds over the central western part of the Greenland ice sheet are optically thin and high, and the cloudiness is relatively small. This is in sharp contrast with Vatnajökull where low clouds are present most of the time and lead to a very strong increase in global radiation with altitude.

The data point from the Pasterze marked * is an outlier caused by intense reflection from surrounding slopes and should not be used to determine an altitudinal gradient [13]. The variability for the Vatnajökull data is larger than for the other two data sets. This is due to the fact that the stations were widely spaced, covering the entire ice cap (area: 8400 km^2).

Clearly, all data sets show a consistent positive gradient in the global radiation with altitude. Unfortunately, it is very hard to determine the relative importance of the processes discussed above. For the Pasterze the decrease in shading when going upwards plays an important role [13]; this is not the case for

Vatnajökull and West Greenland. In principle one could learn more by looking at the characteristic patterns of global radiation for totally clear days. Unfortunately, these are very rare, especially on Vatnajökull.

It is therefore difficult to model the atmospheric transmissivity over glaciers from the meteorological data sets available. The best one can do is to lump everything together in τ_a and use a simple parameterisation based on far more extensive measurements (but not on glaciers...). It is clear that solar elevation has to appear in such a parameterisation, since the effective optical mass depends on it (decreasing mass when the sun gets higher). The same applies to the altitude of the site considered. Among the various expressions that have been proposed in the literature let us look at:

$$\tau_a = \left(0.79 + 2.4 \cdot 10^{-5}\, h\right)\left(1 - 0.08\,\frac{(\pi/2) - e}{\pi/2}\right). \qquad (3.14)$$

Thus the transmissivity increases with altitude h and solar elevation e (see *figure 3.12*). The maximum difference in transmissivity for a low and high sun is about 7%.

In a way, this concludes a description of how the solar radiation impinging on a glacier surface can be calculated. However, we should remain aware of the simplifications that have been made. As stated earlier, eq. (3.10) is simple because the interactions are not formulated explicitly. As a final illustration we note that multiple reflection between glacier surface and atmosphere is even important in clear-sky conditions. In the absence of geometric factors ($\tau_g = 0$) and clouds the global radiation can be written as

$$Q = \tau_a \left[\frac{1}{1 - \alpha_{air}\,\alpha_s}\right] Q_E \qquad (3.15)$$

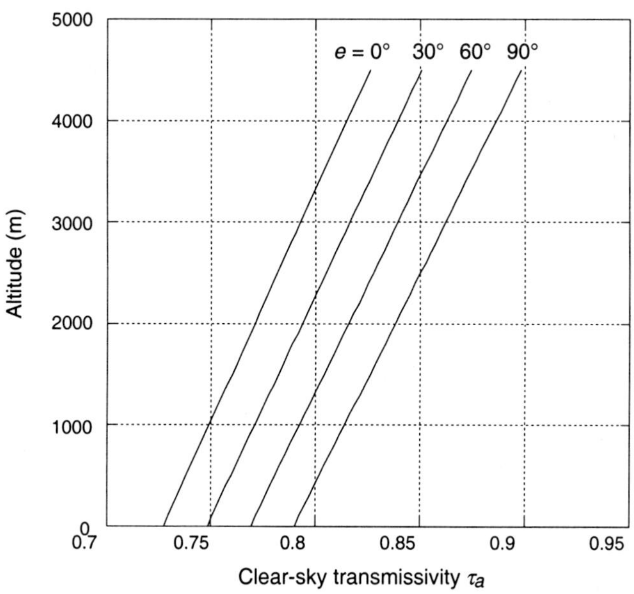

Figure 3.12. Parameterisation of the clear-sky transmissivity according to eq. (3.12) for different values of the solar elevation e.

The term within brackets, containing the surface albedo α_s and the clear-sky albedo α_{air}, is the correction to the atmospheric transmissivity due to multiple reflection. With characteristic values of 0.60 and 0.07 for α_s and α_{air}, it amounts to 1.044. In many studies this correction factor is simply absorbed in the transmissivity, giving a slightly larger value of τ_a. In conditions with high surface albedo and low stratus clouds the correction will be much larger. Since the surface albedo normally increases when going up-glacier, the increase in global radiation will be enhanced by the process of multiple reflection. This may explain a significant part of the altitudinal gradients seen in *figure 3.11*.

3.4 The surface albedo

In Chapter 4 we will see that solar radiation contributes most to the surface energy flux when melting occurs. This makes the surface albedo a particular important quantity deserving a great deal of attention. Albedo depends in a complicated way on crystal structure, surface morphology, dust and soot concentrations, morainic material, the presence of liquid water in veins and at the surface, solar elevation, cloudiness, etc. It appears to be a fairly hopeless task to quantify and model all these factors, but maybe the most important can be captured by looking carefully at the sparse data available.

The spatial and temporal variation of the albedo is large. This can be observed when crossing glaciers on foot, but is far clearer when one looks at satellite images of sufficiently high resolution. As an example we consider four Landsat-5 images of the Haute Glacier d'Arolla analysed by Knap et al. [16]. The Haute Glacier d'Arolla is a small glacier in Valais, Switzerland. It is about 4 km long, has an area of 6.3 km^2, and covers an altitudinal range from 2600 to about 3500 m. The Landsat Thematic Mapper images are quite useful because the resolution (30 m) is good enough to map valley glaciers. The disadvantage is the long repeat time of 16 days. Since cloud-free images are required to determine the surface albedo from the satellite measurements, one is lucky if one gets several useful scenes within a melt season. In the summer of 1993 this was the case for some glacierised parts of the Alps. The Landsat overpasses on May 29, August 1 and 17, and September 2 produced high-quality scenes. In the same summer many ground-based measurements of surface albedo were carried out, creating a unique opportunity for comparing satellite-derived albedo with in situ observations. *Figure 3.13* shows the distribution of the surface albedo calculated from the satellite data. This is not a trivial matter because it involves the conversion of band albedos to the spectrally integrated albedo (Landsat provides radiances in distinct parts of the spectrum) and some assumptions about the nature of the reflected radiation [16].

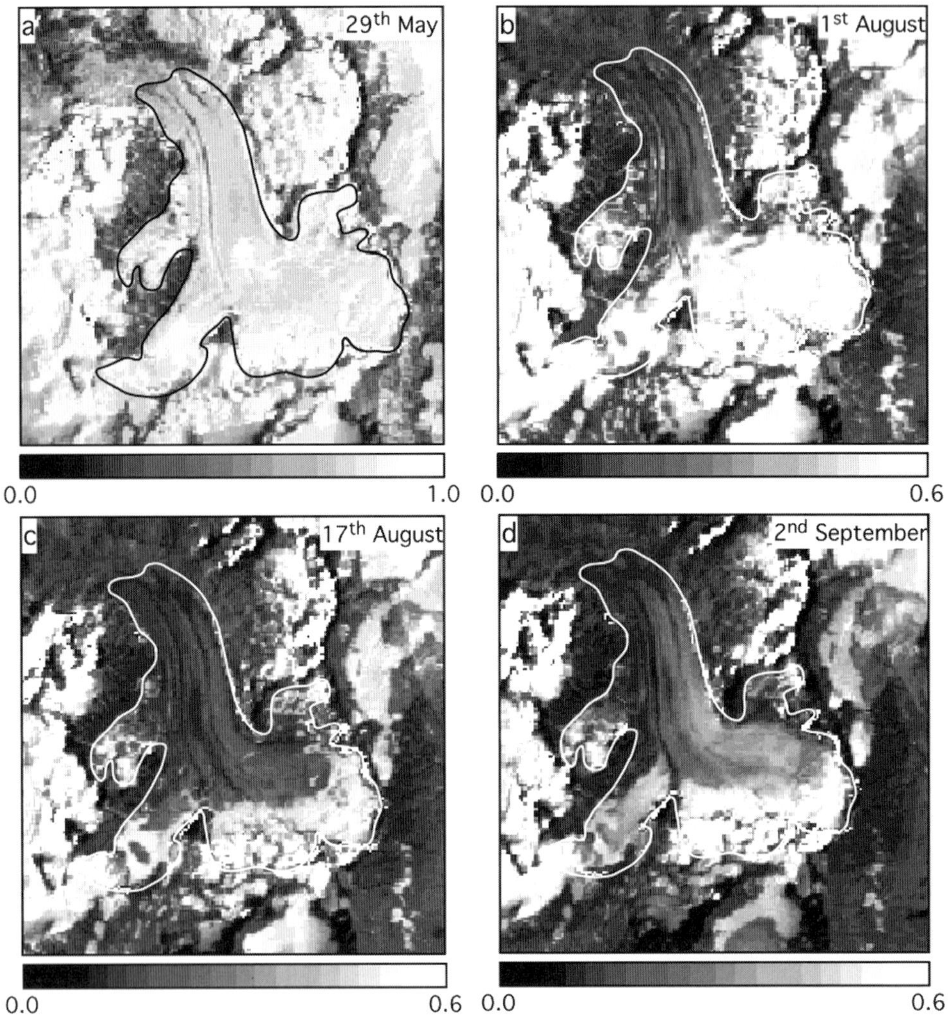

Figure 3.13. The surface albedo of the Haute Glacier d'Arolla as calculated from Landsat 5 Thematic Mapper data for 1993 [16]. The contour shows the approximate outline of the glacier. Note that the albedo scale for the first picture (29th May) runs from 0 to 1 and for the other pictures from 0 to 0.6.

On 29th May the glacier is almost entirely covered by snow, but the medial moraines show up as darker lines. On 1st August the lower part of the glacier is free of snow. The albedo is rather low and longitudinally oriented structures, representing bands of debris, are apparent. The snow line is at an altitude of about 2850 m. By 17th August it is at an altitude of 3000 m. The clear distinction between a bare ice zone with (very) low albedo (mean value 0.13) and a snow zone with high albedo (mean value 0.55) fits the anticipated picture well. However, the interpretation is not always simple. The fourth scene in *figure 3.13* (2nd September) shows a gradual increase in albedo when going up the glacier. Analysis of the meteorological data revealed that about 10 cm of fresh snow fell on the glacier on 28th August. Although most of this snow melted away within a few days, enough patches were left to make the albedo, detected by the Landsat Thematic Mapper, rather high. However, there may also be a bias in the retrieval algorithm associated with the lower solar elevation.

Satellite images provide the best data for studying the spatial patterns, but quantifying the processes that cause temporal changes requires continuous measurements of the albedo in conjunction with other meteorological observations. First of all albedo should be defined more precisely. The term albedo is normally used when the ratio of outgoing and incoming solar flux through a horizontal surface is considered. We can use the term parallel albedo when this surface is taken parallel to the local glacier surface. The concept of parallel albedo is sometimes useful, but there are potential problems with the interpretation when the sun is very low [17].

In fact, a low sun is always problematical because the fluxes ultimately become so small that they are of the same order of magnitude as the measurement errors. Then albedo can take on more or less any value. Another source of error is the orientation of the pyranometers. Because a melting surface changes shape all the time, pyranometers mounted on masts do not remain perfectly horizontal. Corrections can be made for any tilt, but this requires an assumption about the ratio of diffusive radiation to direct radiation. If data from cloudless days show a clear asymmetry in albedo values for the morning and afternoon, this normally points to a tilt in east-west direction. But again one has to be careful: other factors, like the absence or presence of melt water at the surface, may cause a difference in the albedo for the morning and the afternoon.

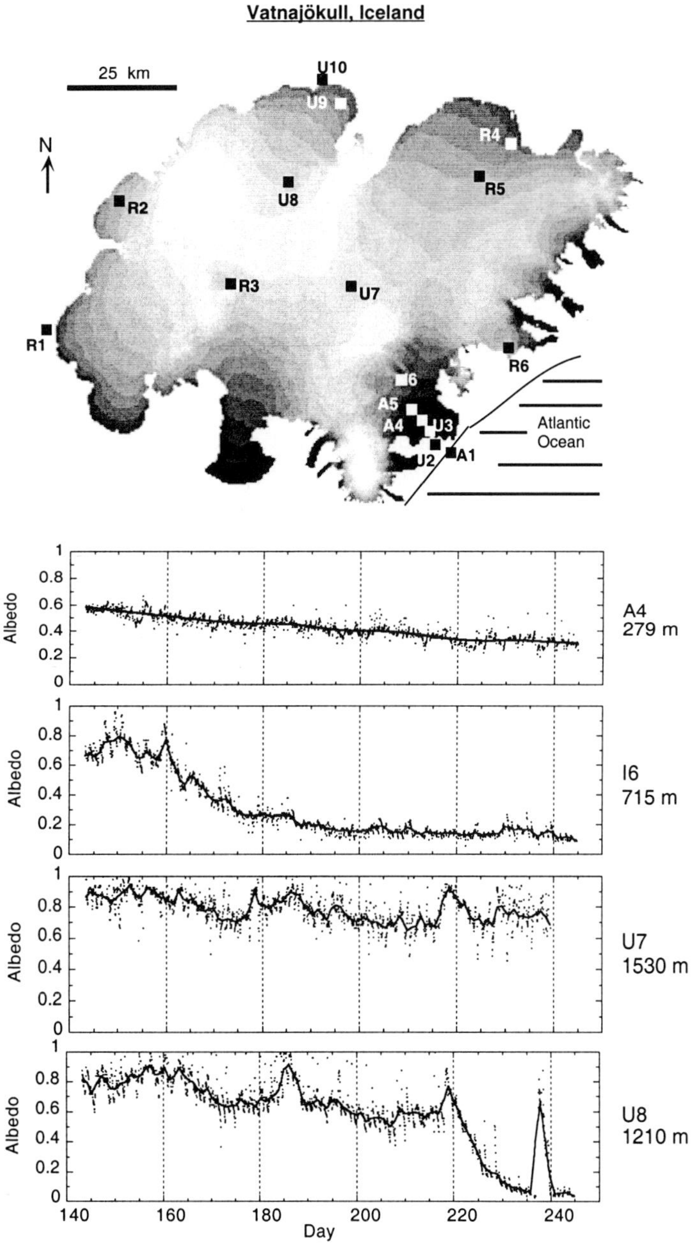

Figure 3.14. A selection of albedo measurements from VATEX-96. The map shows the orography of Vatnajökull, ranging from close to sea level (black) to about 2000 m (white). The weather stations that were in operation during the experiment are all indicated. The graphs show hourly albedo values together with a smoothly fitted curve for four stations. Day 140 corresponds to 19th May, day 250 to 5th September.

Normally some time averaging will be applied before data are analysed. This should be reflected in the wording, e.g. the hourly albedo is defined as

$$\alpha_{hour} = \frac{\overline{Q_r}}{\overline{Q}} \neq \overline{\left(\frac{Q_r}{Q}\right)}, \tag{3.16}$$

Here Q_r is the reflected radiation and Q the global radiation. The overbar indicates the mean value over one hour. The hourly albedo therefore is the ratio of hourly mean values of reflected and global radiation - it differs from the hourly mean albedo! Similarly, the daily albedo differs from the daily mean albedo. Daily albedo is a much more useful quantity than daily mean albedo, because the latter is hard to define for periods with low sun or during the night.

Figure 3.14 shows a selection of measurements from the field campaign on Vatnajökull (summer 1996). The dots represent values of the hourly albedo. Measurements during nighttime have been excluded. There is still some scatter in the data due to hours with low values of global radiation. Another source of noise is precipitation, notably snowfall. When snow adheres to the upward-facing sensor and not to the downward-facing sensor albedo values exceeding one can occur. These do not show up in the figures because the scale is cut off at 1.

In spite of these irregularities the typical evolution of the albedo in the course of the melt season emerges nicely. Station A4 was on a relatively clean part of the glacier, and the disappearance of the snow is not seen as a clear step. Instead, there is general trend from 0.6 at the end of May to about 0.3 in late summer. At I6 the situation is totally different. By the time the snow has disappeared a very dirty ice surface remains with an albedo of about 0.15. Such low values are not uncommon on Vatnajökull because of the large amounts of volcanic ash layers accumulating at the surface. At U8, which is at a relatively high latitude, the ice does not surface until the middle of August. The albedo then drops to even below 0.1 and the surface is effectively black. After a snowfall event on day 237 the black surface appears again. At the highest station in this selection, U7, there is a snow surface the whole summer, although significant melt took place. The albedo varies between 0.6 and 0.95, depending on the age of the snow. Because of the relatively high air temperature, metamorphosis of fresh snow is fast and the albedo drops rapidly after a snowfall event.

We go on to look at more data. *Figure 3.15* shows a 1-year record of daily albedo from the weather station on the Morteratschgletscher [12]. Snow depth is also shown, because it explains most of the albedo variations seen. The snow depth was measured with an ultrasonic ranger (20 kHz) which sampled every 3 hours. It is clear that a small amount of fresh snow brings the albedo up to about 0.8. Many peaks in the albedo record can be linked to the snow depth record. In the course of spring the snow albedo tends to decrease more or less steadily until day 215, when 20 cm of snow fell.

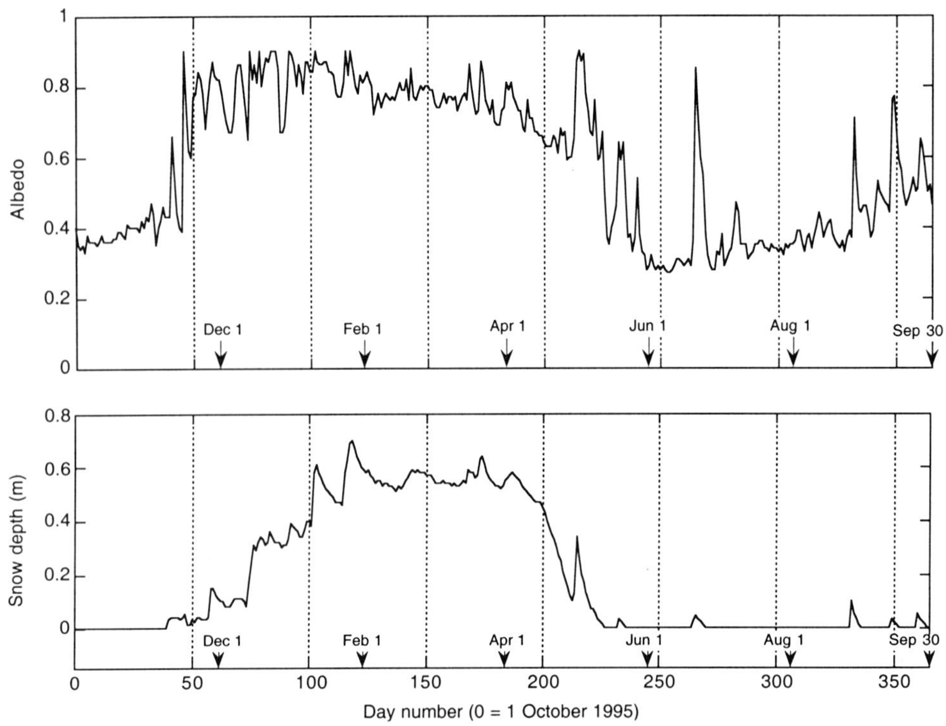

Figure 3.15. A 1-year record of daily albedo from the automatic weather station located on the Morteratschgletscher, Switzerland (see *figure 3.7*). The snow depth was measured with an ultrasonic ranger.

The ice albedo is between 0.3 and 0.4 but increases during the last part of the ablation season.

A scatter plot of all half-hourly values of global and reflected radiation for this year shows a clustering of data points (*figure 3.16*). The points are aligned along a characteristic albedo of snow (~0.7) and a characteristic albedo of ice (~0.3). For some data points the reflected radiation exceeds the global radiation. These are the cases where the upward-facing sensor is (partly) covered by snow.

The annual albedo is 0.491 (the annual mean albedo is 0.555). Therefore at this site about half of the global radiation is actually absorbed by the glacier surface.

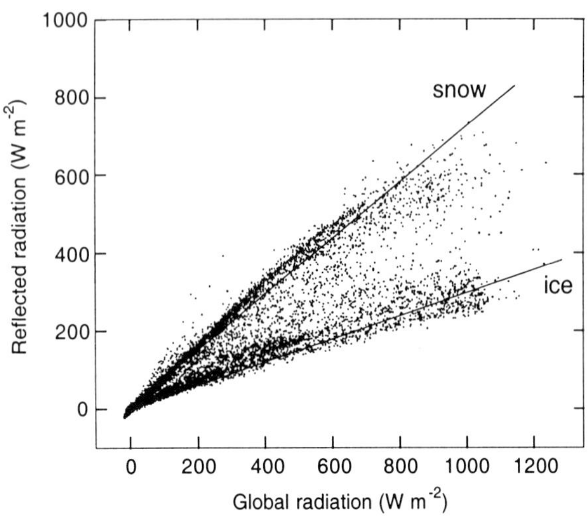

Figure 3.16 One year of half-hourly values of global and reflected radiation from the Morteratsch station.

Recalling that the global radiation was 49% of the extra-terrestrial irradiance we conclude that only 24% of the solar radiation available at the top of the atmosphere is absorbed at the glacier surface.

3.5 Longwave radiation

As for shortwave radiation, the basic theory for calculating the longwave fluxes for given profiles of atmospheric composition, clouds and temperature is well developed. However, these profiles are seldom measured over glaciers! One can either assume standard profiles for most of the constituents and run advanced radiation codes or one can rely on simpler schemes that have been calibrated carefully against measurements.

Do such measurements exist for glaciers? Yes they do, but the data sets are really limited. Measurements of longwave radiation have been made as part of most of the energy-balance studies mentioned earlier. Unfortunately, it is not easy to measure the upward and downward longwave fluxes accurately. The performance of the instruments depends on temperature fluctuations and in particular on the presence of droplets, rime or snow on the sensor. Nevertheless, some useful data exist for a few sites in summer.

To illustrate the relation of longwave radiation to other meteorological quantities, *figure 3.17* shows a record of longwave radiation with global radiation, relative humidity and air temperature. Again, the data are from the station on the Morteratschgletscher. Although cloud observations are not available, a good impression can be obtained from the global radiation. We can be fairly sure that 3rd August was a cloudy day, 4th and 5th were partly cloudy, and 6th and 7th were bright days with no or very little cloud cover.

The outgoing longwave radiation is rather constant as one would expect when the surface is melting all the time. However, the flux is larger than that emitted by a black body at melting point, which is 315 W m^{-2}. The difference can be attributed to the fact that the sensor is 3.5 m above the surface. The air between the surface and the sensor has a temperature of typically 10°C and will make a positive contribution to the upward flux measured at 3.5 m.

The incoming longwave radiation shows more pronounced variations. In the first half of the record this is due first of all to cloudiness and humidity, whereas in the second half it is mainly a response to air temperature.

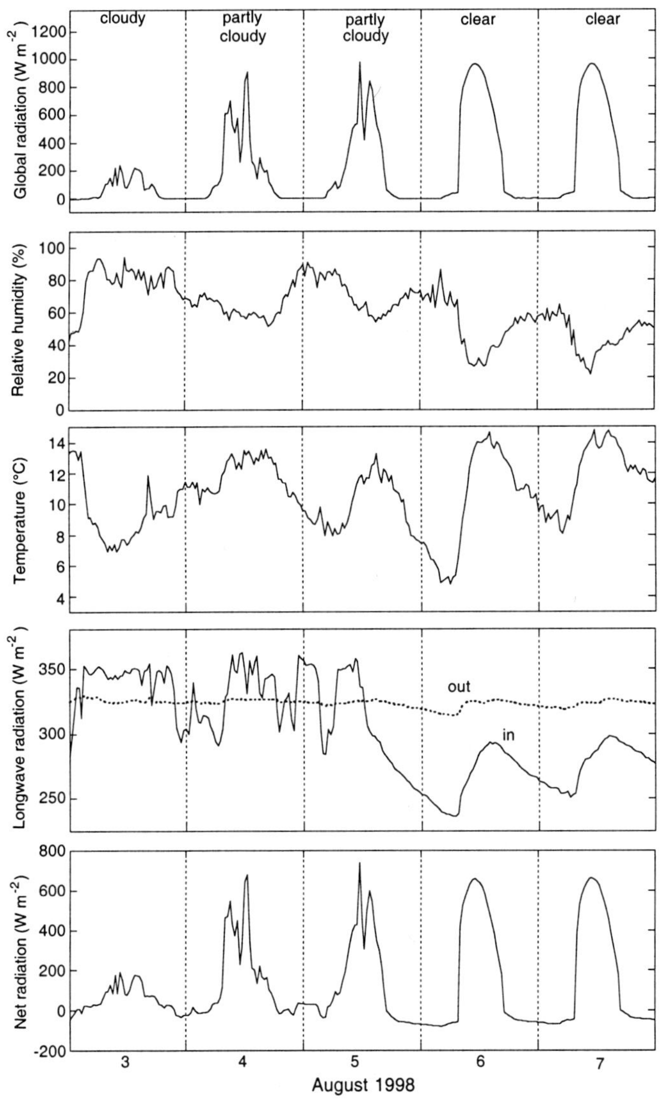

Figure 3.17. Measurements from the Morteratsch AWS, showing incoming and outgoing longwave radiation in relation to other meteorological quantities (half-hourly mean values.) Note the differences in vertical scale on the radiation panels.

The net radiation is shown at the bottom of the figure. The global radiation clearly dominates the picture, in spite of the fact that part of it (~30%, not shown) is reflected. However, from this we cannot conclude that longwave radiation is insignificant. It forms a larger part of the energy budget in autumn and winter. In addition, we should realise that air temperature has a direct effect on the longwave radiation - not on the solar radiation. Longwave radiation thus plays an important role as far as the climate sensitivity of glacier melting is concerned.

Greuell et al. [13] used the PASTEX-94 data to derive a parameterisation for the incoming longwave radiation. They started by correcting for a contribution from the surrounding slopes. This contribution can be significant locally, but is probably less important when I_{in} is considered for larger parts of a glacier.

The basic assumption now is that the contributions from the cloudy and non-cloudy parts of the sky can be separated [15,19]:

$$I_{in} = \left[\varepsilon_{cs}(1-n^p) + \varepsilon_{cl}\, n^p\right] \sigma T_a^4 \,. \tag{3.16}$$

Here ε_{cs} and ε_{cl} are the characteristic emissivities for clear-sky and fully cloudy conditions, respectively. Cloudiness is denoted by n and air temperature at 2 m by T_a. The exponent p is determined by fitting eq. (3.16) to measurements. The clear-sky emissivity depends on the concentration of greenhouse gases, all lumped together in a constant, and on vapour pressure e_a and temperature (both at 2 m):

$$\varepsilon_{cs} = 0.23 + b\left(\frac{e_a}{T_a}\right)^{1/8} . \tag{3.17}$$

Again, the parameter b has to be found by fitting eq. (3.17) to observations. Greuell et al. carried out the analysis for two stations on the Pasterze: station U2 (2310 m) on the glacier tongue and station U5 (3225 m) in the upper accumulation zone. After correcting for the (small) amount of longwave radiation coming from the surrounding slopes, they found as best parameter values: p = 2, ε_{cl} = 0.976, b = 0.475 (at U2), b = 0.407 (at U5). They attribute the difference in b to the fact that during melt conditions the temperature inversion just above the glacier surface (see *figure 3.1*) decreases with altitude. Thus, for a given value of T_a at 2 m the mean air temperature in the lower tens of metres will be larger for the higher station. This provides a plausible explanation for the decrease of b with altitude.

The parameterisation described above has to be used with care, especially with regard to input data. When temperature and humidity are measured at a different height the parameter values have to be changed. In practice it is preferable to reduce the measurements to the level for which the parameterisation was derived, but from a general point of view it would be better to express the incoming longwave flux in terms of temperature and humidity at a greater height above

the surface, preferably above the inversion. However, very few data of this kind are available.
Another weakness of eq. (3.16) is that the cloud contribution to I_{in} does not depend on cloud height and/or temperature. From a physical point of view this would be more desirable [19,20]. However, in practical calculations these quantities are generally not known. By fitting eq. (3.16) to data the effect of cloud base temperature is absorbed in the cloud emissivity. When the cloud base is higher, the value of ε_{cl} will be slightly lower because of a lower temperature at the cloud base.

3.6 The atmospheric boundary layer over glaciers

An atmospheric boundary layer is present everywhere. It is generally defined as the part of the atmosphere that is directly influenced by the surface. This influence derives from the exchange of momentum (friction), heat and moisture at the surface. In contrast to the free atmosphere, the structure of the boundary layer exhibits a marked daily cycle. At high latitudes, however, this is not necessarily the case because tha daily cycle in the surface fluxes can be quite small. For instance, katabatic flows over Antarctica in wintertime do not have a clear daily cycle. Yet these flows should be considered as boundary layer flows in which surface friction and the turbulent sensible heat flux are important components of the momentum and heat budget. In a way the same situation is found over melting glacier surfaces. During a large part of the summer temperature and vapour pressure at the surface have no daily cycle. Therefore the forcing of a glacier boundary layer from below is fairly constant and variations in its structure will be related to what happens higher up.
In *figure 3.18* a typical summer time situation in a glacierised mountain region is sketched. Mountains appears as large roughness elements to the synoptic-scale atmospheric flow and a large-scale boundary layer will be present. In the valleys the diurnal rhythm of the valley/mountain wind system dominates the flow. It interacts with the large-scale boundary layer in a complex way. The glacier wind is a tiny element of this system but nevertheless regulates the turbulent fluxes to or from the glacier surface.
Many balloon ascents have made it clear that the glacier wind is rather shallow (*figure 3.1*), normally only some tens of metres. Measurements have also shown that it does not penetrate very far into the valley. At the point where the valley and glacier wind meet a weak front is present (indicated by F in *figure 3.18*). On really warm summer days the valley wind can become so strong that it erodes the glacier wind on the lower part of the glacier: the front moves upwards. This has been observed on several occasions on Austrian glaciers and on the Morteratschgletscher [22]).

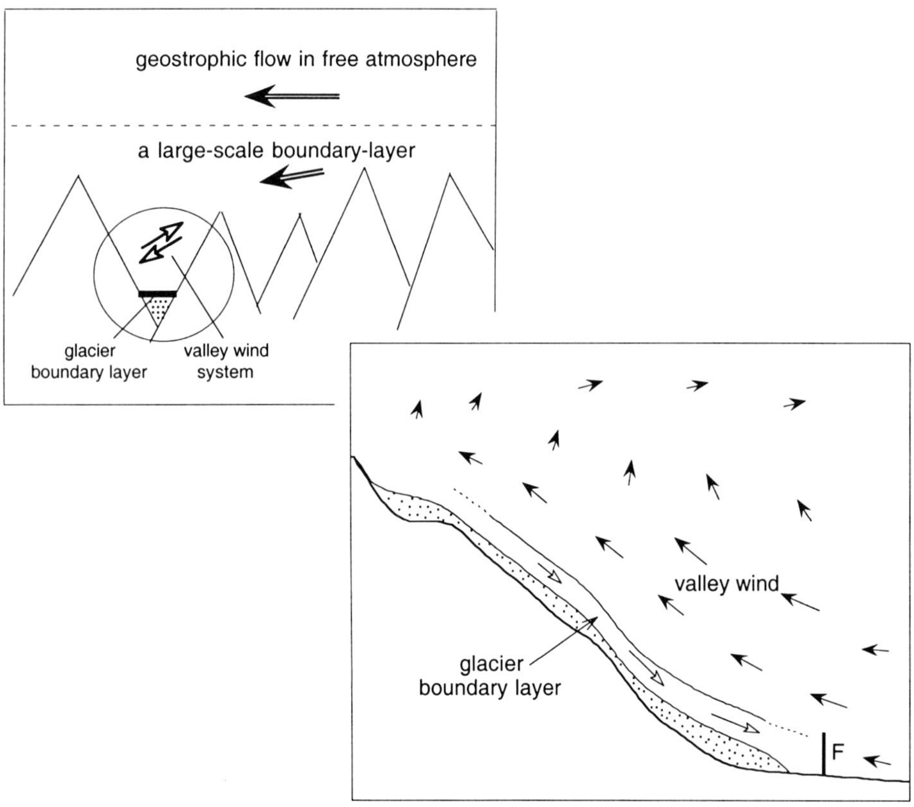

Figure 3.18. The basic structure of atmospheric circulation in a glacierised mountain region.

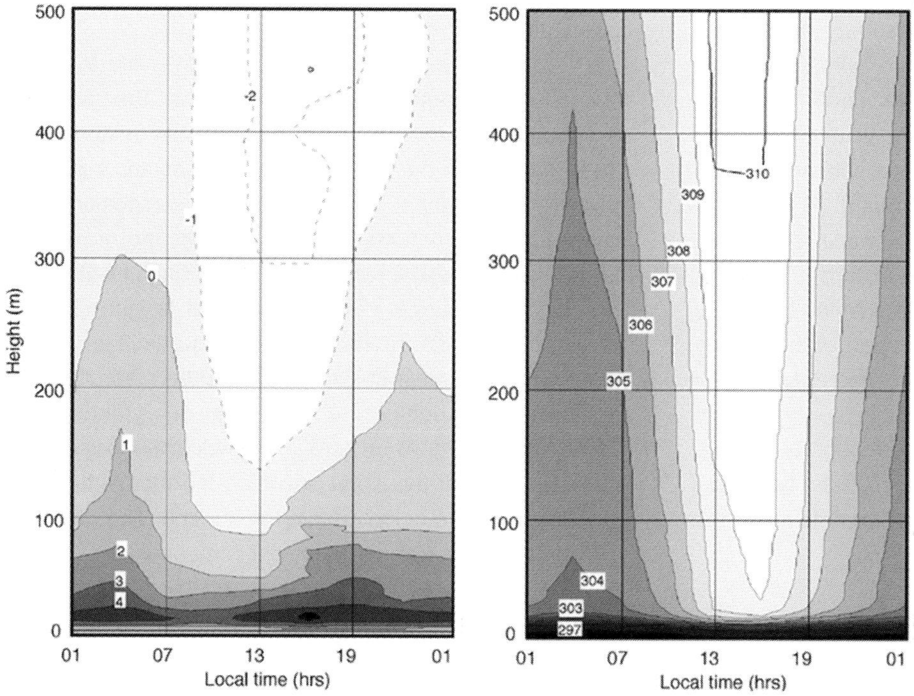

Figure 3.19. Daily cycle of downslope wind speed and potential temperature as measured with balloon soundings during PASTEX. The patterns shown are the mean for a 12-day period in summer with weak synoptic-scale forcing [23].

The temperature and wind profiles obtained by cable balloon during PASTEX-94 form a good data set for studying the glacier / valley wind system. Van den Broeke [23] has made an analysis of these data. *Figure 3.19* provides insight into the observed daily cycle during a period of fair weather with weak synoptic-scale forcing. The measurements were performed on the glacier tongue. Shown are the mean fields over a 12-day period of fair weather and weak synoptic-scale forcing (23rd July until 3rd August 1994).

The katabatic flow exhibits a semi-diurnal cycle, with largest vertical extent (and wind speeds) late at night and in the late afternoon. Van den Broeke suggests that the late night maximum is associated with mass flux towards the glacier from the cooler valley walls. The maximum in the late afternoon is due to the maximum in buoyancy forcing, because at this time the air filling the valley reaches its maximum temperature. The development and decay of the valley wind are clearly seen. The speed of the valley wind is typically 2 m s^{-1} and air temperature at 100 m above the glacier surface has a daily range of the order of 10 K.

So far we have only discussed the situation in summer because it is most relevant to the melt process and has consequences for the specific mass balance. In winter the contrast between glacier and surrounding terrain is much smaller and local circulations are less glacier-specific.

It is interesting to compare wind data from the station on the Morteratschgletscher for summer and winter. *Figure 3.20* shows scatter plots of wind speed and wind direction for the summer and winter half year. It is clear that air flows down the glacier most of the time.

An attempt to relate the wind direction to the large-scale synoptic circulation yields very small correlation coefficients. In the few cases when strong northerly winds are observed the large-scale circulation tends to be northerly too.

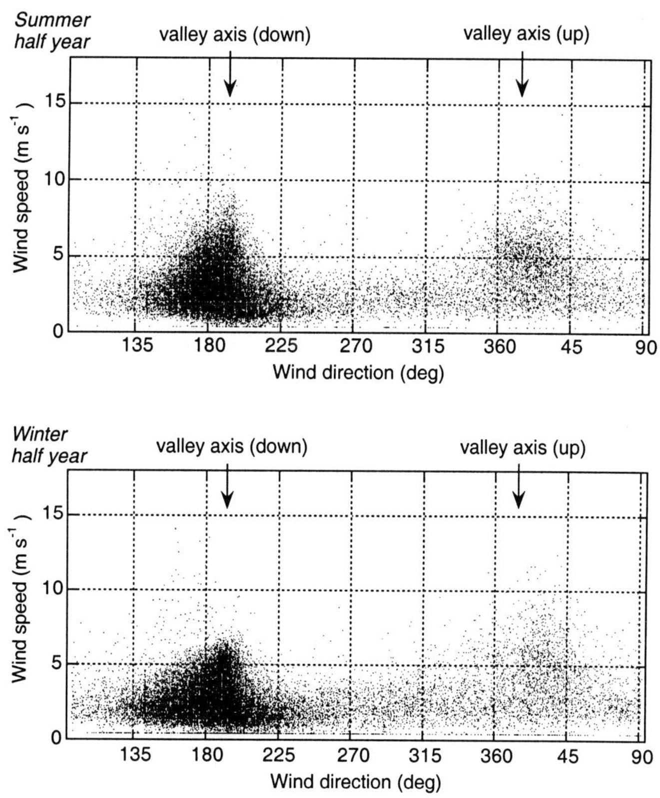

Figure 3.20. A plot of wind speed versus wind direction from the Morteratsch station. The summer half year is defined as 1st May - 30th October, the winter half year as 1st November - 30th April. Data are for the period 1st October 1995 until 30th November 1998. Each plot contains 26304 data points (half hourly values).

What we see in *figure 3.20* are katabatic flows driven by a downward sensible heat flux at the glacier surface. In summer the heat is used mainly for melting, in winter it partly compensates for the strong radiative cooling. When comparing the two plots we should bear in mind that the average height of the sensors is greater in summer (3.5 m) than in winter (3.5 m minus the snow depth). The slightly smaller average speed of katabatic flow in winter suggested by *figure 3.20* may be due to the difference in sensor height.

Turbulent fluxes play an important role in shaping boundary layers. They also account for a significant part of the energy transfer between glacier surface and atmosphere. In view of this we will consider the theory of turbulent transfer in the next section. The discussion will be of an introductory nature. There are many textbooks about the atmospheric boundary layer and turbulence which give thorough treatments of the theory [e.g. 24].

3.7 Turbulence and boundary-layer profiles

Many natural flows are turbulent rather than laminar. Turbulent motion is irregular and therefore hard to describe in mathematical terms. One way that has been tried is known as Reynolds decomposition. The value of a fluctuating quantity X is written as the sum of a time mean value (overbar) and a perturbation (prime) from this mean:

$$X = \overline{X} + X' \ , \qquad (3.18)$$

$$\overline{X'} = 0 \ . \qquad (3.19)$$

The product of two fluctuating quantities represents a flux. For instance, denoting the horizontal and vertical components of the wind vector by u and w, the upward or downward flux of horizontal momentum averaged over some time can be written as $\overline{uw}$ [note that this can be interpreted equally well as the horizontal flux of vertical momentum!]. Inserting eq. (3.19) we find:

$$\overline{uw} = \overline{(\overline{u} + u')(\overline{w} + w')} = \overline{u}\,\overline{w} + \overline{u'w'} \ . \qquad (3.20)$$

The total flux therefore consists of two contributions: one from the mean fields and one from the turbulence. Close to a horizontal surface $\overline{w}$ will vanish, so only the turbulent part remains. The turbulent flux can only be non-zero when there is a correlation between the quantities, so the flow cannot be totally random.

Similarly, the vertical turbulent heat flux can be written as

$$H_s = \overline{w'T'} \ . \qquad (3.21)$$

When, on average, warm air moves upwards and cold air downwards the net heat flux is upwards. In glaciology one tends to define the sign in the opposite way, i.e. to define a downward heat flux as positive, see eq. (3.1).

The theory of the turbulent boundary layer starts with the Navier-Stokes equations that describe the flow of a continuum. Normally a Cartesian coordinate system is used where x and y are the horizontal coordinates and z is pointing upwards. A treatment of a Reynolds decomposition of the Navier-Stokes equations is beyond the scope of this book. Instead, we will make a few large jumps through the theory.

The momentum budget varies strongly with height. Sufficiently close to the surface the frictional force is dominant (say at a height d_f). In the case of a steady state the mean flow is not accelerated or decelerated, so the divergence of the momentum flux should be zero. This implies the existence of a so-called constant flux layer, in which the momentum flux is independent of the height above the surface. The momentum flux F_u has a molecular and turbulent component:

$$F_u = -\nu \frac{d\overline{u}}{dz} + \overline{u'w'} \ . \qquad (3.22)$$

The molecular component is proportional to the velocity gradient and the dynamic viscosity ν.

In K-theory it is assumed that turbulent transport is of a diffusive nature, implying that fluxes can be expressed as eddy viscosity multiplied by the gradient in the mean flow. For example

$$\overline{u'w'} = -K_m \frac{d\overline{u}}{dz} \quad (= \tau/\rho = -u_*^2) \ , \qquad (3.23)$$

where K_m is the eddy diffusivity for momentum. Eq. (3.23) also defines the friction velocity u_*, which is just the square root of the momentum flux. It is sometimes referred to as the turbulent velocity scale. By definition, the momentum flux equals the shear stress τ divided by the air density ρ. Note that ν is a property of the fluid, K_m is a property of the flow!

Turbulent eddies are more effective in smoothing gradients when they are larger. Close to the surface the turbulent eddies must decrease in size. Very close to the surface there is no turbulence and molecular processes control the momentum flux. This is the viscous layer. Outside this layer, but inside the boundary layer, the turbulent eddies dominate and the turbulent stress is much larger than the viscous stress.

So how can we find K_m? An expression can be found with a dimensional argument. The only relevant quantities are the height z and the wind shear $d\overline{u}/dz$. Because K_m has the unit $m^2\ s^{-1}$, it follows that

$$K_m = (\kappa z)^2 \frac{d\overline{u}}{dz} \ , \qquad (3.24)$$

where κ is the Von Kárman constant (an experimental constant normally taken as 0.4). A physical interpretation of eq. (3.24) is that κz is the mixing length and κz $(d\overline{u}/dz)$ the characteristic momentum difference experienced by the

eddies over the mixing length.
Combining eqs. (3.23) and (3.24) we find

$$\frac{d\bar{u}}{dz} = \frac{u_*}{\kappa z} , \qquad (3.25)$$

which can be integrated to give:

$$\bar{u} - \bar{u}_0 = \frac{u_*}{\kappa} \ln(z/z_0) . \qquad (3.26)$$

Two integration constants appear. We take $\bar{u}_0 = 0$, which defines the roughness length z_0 (the height at which the wind speed in the logarithmic profile equals zero). So with $\bar{u}_0 = 0$ we obtain the well-known logarithmic wind profile:

$$\bar{u}(z) = \frac{u_*}{\kappa} \ln(z/z_0) . \qquad (3.27)$$

Figure 3.21 shows wind profiles measured during PASTEX-94 in the case of a wind blowing up-glacier. This does not happen too often as the katabatic flow normally dominates. However, this sample is shown to avoid the complications that can arise if there is katabatic flow. Each profile represents a half-hourly mean state. The profiles are plotted with a linear and a logarithmic height scale.
The profiles do indeed tend to be more or less linear on the logarithmic scale. A dashed line is drawn to characterise the mean state. The intersection of this line with the height-axis gives the roughness length z_0, whereas the friction velocity u_* and hence the momentum flux is determined by the slope.
It is clear from the picture that the roughness length is generally not well defined. Making linear fits to the six profiles shown would give a large scatter, in spite of the fact that the half-hourly mean profiles are in a sequence and cover no more than 3 hours. Values of z_0 over glaciers reported in the literature range from 0.1 to 100 mm.
Now we turn to heat and water vapour. By analogy with eq. (3.23) we have (taking account of the fact that the eddy diffusivities may be different):

$$\overline{w'T'} = -K_{se} \frac{d\bar{T}}{dz} \quad (\equiv -u_* T_*) , \qquad (3.28)$$

$$\overline{w'q'} = -K_{la} \frac{d\bar{q}}{dz} \quad (\equiv -u_* q_*) . \qquad (3.29)$$

These expressions also define the turbulent scales for heat (T_*) and for the water vapour mixing ratio (q_*).
The dimensional argument leading to eq. (3.25) can be applied again, yielding:

$$\frac{d\bar{T}}{dz} = \frac{K_{se}}{K_m} \frac{T_*}{\kappa z} , \qquad (3.30)$$

and

$$\frac{d\bar{q}}{dz} = \frac{K_{la}}{K_m} \frac{q_*}{\kappa z} . \qquad (3.31)$$

The corresponding logarithmic temperature and humidity profiles are

$$\bar{T} - \bar{T}_0 = \frac{K_{se}}{K_m} \frac{T_*}{\kappa} \ln(z/z_T) , \qquad (3.32)$$

$$\bar{q} - \bar{q}_0 = \frac{K_{la}}{K_m} \frac{q_*}{\kappa} \ln(z/z_q) . \qquad (3.33)$$

Here $\bar{T}_0$ and $\bar{q}_0$ are the (near-)surface values. The roughness lengths for temperature and humidity are denoted by z_T and z_q; they generally differ from z_0.
The profile relations derived above are valid only for near-neutral conditions, i.e. when the potential temperature changes little with height. In the glacier boundary layer this is rarely the case and we should realise that the theory summarised above provides no more than a framework for further considerations.

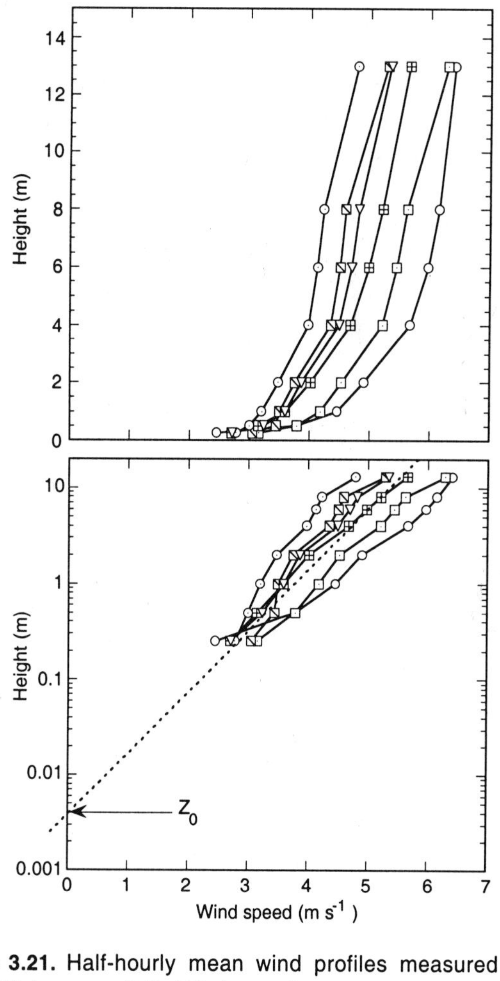

Figure 3.21. Half-hourly mean wind profiles measured during PASTEX (summer '94). Wind speed was measured at 8 levels. The profiles are plotted with a linear and logarithmic scale for height. The dashed line shows an attempt to find a typical roughness length.

3.8 Turbulent fluxes

Building on the formulations of the previous section, we can write the turbulent fluxes of momentum, sensible heat and latent heat as:

$$\tau = -\rho\,\overline{u'w'} = \rho K_m \frac{d\bar{u}}{dz} = \rho u_*^2 \ , \tag{3.34}$$

$$H_{se} = -\rho c_p \overline{w'T'} = \rho c_p K_{se} \frac{d\bar{T}}{dz} = \rho c_p u_* T_* \ , \tag{3.35}$$

$$H_{la} = -\rho L_v \overline{w'q'} = \rho L_v K_{la} \frac{d\bar{q}}{dz} = \rho L_v u_* q_* \ . \tag{3.36}$$

How can these fluxes be determined from measurements of wind speed, temperature and humidity? In many applications the turbulent fluxes are calculated with bulk formulas derived from eqs. (3.34)-(3.36) and the profile relations. In the following we drop the overbar to indicate mean values.
Substituting the profiles (3.27) and (3.32) into eq. (3.33) yields

$$H_{se} = \rho c_p u_* T_* = \rho\ c_p \left[\frac{\kappa^2 (K_m/K_{se})}{\ln(z/z_0)\ \ln(z/z_T)} \right] u(z)\,(T(z) - T_0) \ . \tag{3.37}$$

If we now choose a reference height ($z = z_{ref}$) at which measurements are made, for instance 2 m, the expression in square brackets can be defined as the exchange coefficient C_{se}. Therefore

$$H_{se} = \rho c_p C_{se} u\,(T - T_0) \ . \tag{3.38}$$

Similarly, the turbulent fluxes of latent heat and momentum can be written as

$$H_{la} = \rho L_v C_{la} u\,(q - q_0) \ , \tag{3.39}$$

and

$$\tau = \rho C_d u^2 \ . \tag{3.40}$$

L_v is the latent heat of vaporisation of water. The constant C_d is known as the drag coefficient. It is important to remember that the constants C_{se}, C_{la} and C_d depend on the height at which the measurements are made. In fact, they should be considered as empirical constants that depend on the nature of the turbulence and, consequently, on the general atmospheric conditions and the roughness. Values found from field experiments are in the 10^{-2} to 10^{-4} range.
When the atmosphere is not neutrally stratified but is stably stratified (potential temperature increases with height) or unstably stratified (potential temperature decreases with height), the profile relations have to be modified. This can be done with the Monin-Obukov similarity theory (MO-theory [24]), in which profiles and turbulent exchange are governed by the Monin-Obukov length scale L_{MO}. Profiles are then formulated in terms of z/L_{MO}. L_{MO} is determined by the wind shear and the vertical temperature gradient. A treatment of the Monin-Obukov similarity theory is beyond the scope of this book. With regard to the turbulent fluxes of heat and water vapour the major implication is that fluxes are suppressed in stable conditions and enlarged in unstable conditions. Since the temperature and wind profiles themselves depend on the fluxes, it is now more complicated to determine the fluxes from measurements (Munro and Davies [3]). In the first place, measurements are needed from at least two levels. Secondly, some kind of iterative procedure has to be used.
Over melting glacier surfaces, the application of MO-theory turns out to be problematical [25]. Because of the large stability (temperature gradients of the order of 1 K m^{-1}) and the rough surface there is a zero-referencing problem: the fluxes and surface roughnesses determined by MO-theory become very dependent on how the height of the instruments is defined (*figure 3.22*). This is an undesirable situation.
It appears that the use of bulk equations in combination with a method for determining the surface roughness with a microtopographic method yield satisfactory results, and our discussion will be continued along these lines.
After some algebraic manipulation it can be shown that the exchange coefficients are related according to the following equations [27]:

$$C_d = \frac{\kappa^2}{[\ln(z_{ref}/z_0)]^2} \ , \tag{3.41}$$

$$C_{se} = \frac{(K_{se}/K_m)\,\kappa^2 C_d^{1/2}}{\kappa C_d^{-1/2} - \ln(z_T/z_0)} \ , \tag{3.42}$$

$$C_{la} = \frac{(K_{la}/K_m)\,\kappa^2 C_d^{1/2}}{\kappa C_d^{-1/2} - \ln(z_q/z_0)} \ . \tag{3.43}$$

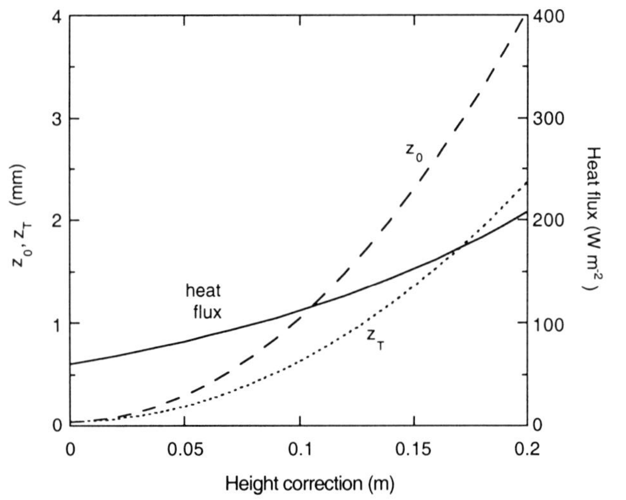

Figure 3.22. Redrawn from Munro's study on Peyto Glacier [25]. The figure shows the effect of height correction on the calculation of roughness length and surface fluxes from profile analysis, and illustrates the zero-reference problem.

where the diffusivity is

$$D=(\rho g)^3 H^3 \left(\frac{\partial h}{\partial x}\right)^2 (f_d H^2 + f_s) \quad . \tag{5.32}$$

As in the previous sections, the height of the bed is denoted by b (so $h = b + H$).

Formally, eq. (5.31) is not a pure diffusion equation because kinematic waves can be a solution [12]. However, the response of a glacier's geometry to a changing mass balance is dominated by diffusion, not by wave propagation as has been suggested in earlier studies [13].

The nonlinear character of the diffusion process is obvious: D increases strongly with slope and ice thickness. A cubic metre of ice in a cold room will not change shape for a very long time. A cubic kilometre of ice, however, would look quite different after a few years!

On a given domain, eq. (5.31) can be solved with standard numerical methods for parabolic equations. Defining the domain requires some care, however. First of all, eq. (5.31) is not valid at the glacier front, because, when ice thickness vanishes, D=0. This would imply that the front cannot move, which is clearly unrealistic. One way to solve this is to use a prescribed snout profile, based on some theory (e.g. [14]), match this with the numerically computed ice thickness sufficiently far away from the glacier front, and determine the actual position of the front from the overall mass budget. The author has made several attempts in this direction, but none of the methods has been entirely satisfactory. Instead, the simpler procedure described below appears to work well.

When the equations are put into incremental form, for instance by straightforward central differencing, the inherent truncation error causes D to have a positive value at the grid point in front of the glacier snout. As illustrated in *figure 5.9*, a staggered grid can be used. D is first evaluated on the regular points (crosses in the figure), and then the term

$$F = D \frac{\partial (b+H)}{\partial x} \tag{5.33}$$

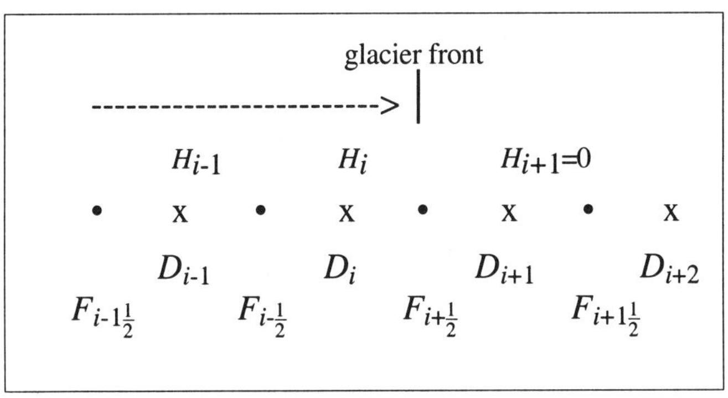

Figure 5.9. Staggered grid to solve eq. (5.31) numerically. The part of the grid around the glacier front is shown.

is evaluated at the staggered points. Adjacent values for *D* are then averaged, a procedure that implies some smoothing and provides numerical stability without additional filtering.

It is obvious that, when the last grid point covered by ice is at i (discrete space variable), the mass flux into grid point i+1 is always nonzero. If this flux exceeds the negative mass balance at i+1, ice will build up and the glacier front will advance. The discrete form of eq. (5.31) therefore does allow movement of the snout! The robustness of this scheme can be tested by imposing different boundary conditions. For instance, one could prescribe

$$D_{iedge+1} = D_{iedge} \quad . \tag{5.34}$$

Therefore the gradient in the diffusivity vanishes at the glacier front. It appears that this does not yield significantly different results, because the position of the glacier front is determined almost entirely by the balance gradient. In fact, simple methods work here because the response of the glacier front on a decadal to century time scale is determined by the overall mass budget, not by the detailed dynamics of the snout [15]. Obviously, such a statement cannot be made for calving glaciers.

Boundary conditions at the head of the glacier are best taken as

$$\left[\frac{\partial h}{\partial x}\right]_{x=0} = 0 \quad \text{[i.e. } h_1 = h_2 \text{ on the grid]} \ , \tag{5.35}$$

implying that the influx of ice at the head of the glacier is zero. The head represents either an ice divide or a wall at which the glacier starts to flow.

When using the scheme outlined above with forward time differencing, the CFL-condition should of course be met [16]

$$\Delta t < \frac{(\Delta x)^2}{D} \quad . \tag{5.36}$$

Since the diffusivity is not constant, the critical time step will vary. With a grid-point distance of $\Delta x = 100$ m, a suitable value for many valley glaciers, a time step of $\Delta t = 0.01$ to 0.05 a generally ensures stability for the forward scheme.

Another popular and more elegant method to integrate parabolic equations of the type considered here is the implicit method, in which the difference equations for the grid points are solved simultaneously. Such a method allows a larger time step, but the gain in computational speed is partly offset by the need to invert large matrices. Nevertheless, those wishing to have an efficient scheme are advised to use the implicit method. Explicit time differencing is easier for coding and more flexible when one is playing around with different formulations of boundary conditions. In any case, modern PCs (status 2000) are so fast that any method is good enough to make a simulation for 5000 years within a few minutes.

Any code should be tested on at least two points:

- mass should be conserved
- the solution should be independent of Δx and Δt.

Truncation errors can be significant when gradients in ice thickness are large (normally close to the glacier front). An acceptable scheme should not have an imbalance larger than a few per mille. Such an imbalance has no noticeable effect on simulated front positions. This is because the glacier model described here represents a highly damped system with a large throughput of mass.

Varying the gridpoint distance and time step is a very useful exercise. Results should not be trusted if, for instance, they change when the gridpoint distance is halved. The gridpoint distance needed to resolve adequately undulations in the bed can be found by trial and error. Anyway, it is advisable to use a smoothed bed, because this is implied by the simple theory of ice flow that forms the basis of the numerical model.

References and further reading

1. J. Weertman (1961): Stability of ice-age ice sheets. *Journal of Geophysical Research* **66**, 3783-3792.

   J. Oerlemans (1981): Some basic experiments with a vertically-integrated ice-sheet model. *Tellus* **33**, 1-11.

2. J.W. Greuell (1992): The Hintereisferner: mass-balance reconstruction and numerical modelling of the historical length variations. *Journal of Glaciology* **38** (129), 233-244.

3. Z. Zuo and J. Oerlemans (1997): Numerical modelling of the historic front variation and the future behaviour of the Pasterze glacier. *Annals of Glaciology* **24**, 235-241.

4. J.F. Nye (1965): The flow of a glacier in a channel of rectangular, elliptic or parabolic cross-section. *Journal of Glaciology* **5**, 661-690.

   W.F. Budd and D. Jenssen (1975): Numerical modelling of glacier systems. *IAHS-AISH Publ. No. 104*, 257-291.

   P.D. Kruss (1983): Climatic change in East Africa: numerical modeling from the 100 years of terminus record of Lewis Glacier, Mount Kenya, *Zeitschrift für Gletscherkunde und Glazialgeologie* **19**, 43-60.

   Ph. Huybrechts, P. de Nooze and H. Decleir (1989): Numerical modelling of Glacier d'Argentière and its historic front variations. In: *Glacier Fluctuations and Climatic Change.* (ed.: J. Oerlemans), Reidel, 373-389.

5. W.D. Mean (1976): *Stress and strain.* Springer, New York, 339 pp.

   W.R. Schowalter (1978): *Mechanics of non-Newtonian fluids.* Pergamon, Oxford, 300 pp.

   D.L. Turcotte and G. Schubert (1982): *Geodynamics. Application of continuum physics to geological problems.* John Wiley, New York, 450 pp.

6. W.S.B. Paterson (1994): *The Physics of Glaciers.* Third edition. Pergamon/Elsevier, New York, 380 pp.

   R. LeB. Hooke (1998): *Principles of Glacier Mechanics.* Prentice Hall, New Jersey, 248 pp.

   C.J. van der Veen (1999): *Fundamentals of Glacier Dynamics.* Balkema, Rotterdam, 406 pp.

7. J. Weertman (1961): Stability of ice-age ice sheets. *Journal ofGeophysical Research* **66**, 3783-3792.

   J. Weertman (1974): Stability of the junction of an ice sheet and an ice shelf. *Journal of Glaciology* **13**, 3-11.

   N. Reeh (1982): A plasticity approach to the steady-state shape of a three-dimensional ice sheet. *Journal of Glaciology* **28**, 431-455.

8. J. Oerlemans and C.J. van der Veen (1984): *Ice Sheets and Climate.* Reidel, 217 pp.

9. J.F. Nye (1970): Glacier sliding without cavitation in a linear viscous approximation. *Proceedings of the Royal Society of London*, Series A **315**, 381-403.

   H. Engelhardt, W.D. Harrison and B. Kamb (1978): Basal sliding and conditions at the glacier bed as revealed by bore-hole photography. *Journal of Glaciology* **20**, 469-508.

   J. Weertman (1979): The unsolved general glacier sliding problem. *Journal of Glaciology* **23**, 97-115.

   L. Lliboutry (1979): Local friction laws for glaciers: a critical review and new openings. *Journal of Glaciology* **23**, 67-95.

   A.C. Fowler (1979): A mathematical approach to the theory of glacier sliding. *Journal of Glaciology* **23**, 131-141.

   B. Hallet (1981): Glacial abrasion and sliding: their dependence on the debris concentration in basal ice. *Annals of Glaciology* **2**, 23-28.

   A. Iken (1981): The effect of subglacial water pressure on the sliding velocity of a glacier in an idealized numerical model. *Journal of Glaciology* **27**, 407-421.

   A.C. Fowler (1987): Sliding with cavity formation. *Journal of Glaciology* **33**, 255-267.

10. M.F. Meier and A. Post (1969): What are glacier surges ? *Canadian Journal of Earth Sciences* **6**, 807-817.

A.C. Fowler (1987): A theory of glacier surges. *Journal of Geophysical Research* **9B**, 9111-9120.

11 W.F. Budd, P.L. Keage and N.A. Blundy (1979): Empirical studies of ice sliding. *Journal of Glaciology* **23**, 157-170.

12. M.J. Lighthill and G.B. Whitham (1955): On kinematic waves. *Proceedings of the Royal Society of London*, Series A **229**, 281-345.

J.F. Nye (1963): The response of a glacier to changes in the rate of nourishment and wastage. *Proceedings of the Royal Society of London*, Series A **275**, 87-112.

13. R.S.W. van de Wal and J. Oerlemans (1995): Response of valley glaciers to climate change and kinematic waves: a study with a numerical ice-flow model. *Journal of Glaciology* **41**, 142-152.

14. J.F. Nye (1967): Plasticity solution for a glacier snout. *Journal of Glaciology* **6**, 695-714.

15. T. Jóhanneson, C.F. Raymond and E.D. Waddington (1989): Time-scale for adjustment of glaciers to changes in mass balance. *Journal of Glaciology* **35**, 355-369.

16. G.D. Smith (1978): *Numerical Solution of Partial Differential Equations: Finite Difference Methods* (2nd edition). Clarendon, Oxford, 304 pp.

C.J.A. Fletcher (1988): *Computational Techniques for Fluid Dynamics* - 2 Volumes. Springer, Berlin.

6. Calculations for a glacier of constant width

6.1 Introduction

The numerical model discussed in the previous chapter can be used to simulate real glaciers. Only a few adjustments are needed to accommodate a parameterisation of the three-dimensional geometry. However, it is instructive to use the numerical flow model for very simple geometries, because it sheds light on the basic relations between bed slope, typical basal shear stress, mean ice thickness etc. Such an analysis may also reveal how realistic the simple analytical model discussed in section 5.4 actually is. In this chapter, therefore, calculations are done with a schematic geometry and a linear balance profile. The model will be applied to real glaciers in Chapters 7 and 8.

6.2 A basic experiment

The first step to take with the numerical model is to simulate a glacier on a bed with a constant slope, and full lateral symmetry. To facilitate a comparison with the simple analytical model, the balance gradient is taken to be linear:

$$B = 0.007\,(h - E) \quad . \qquad (6.1)$$

Figure 6.1 shows a typical result. The slope of the bed is 0.1 and the flow parameters have the values mentioned in section 5.6. The solution shown is for $b_0 - E = 600$ m.

Clearly, there is a certain degree of symmetry around the middle of the glacier, except with regard to the sliding velocity. The sliding velocity increases towards the glacier front because of decreasing ice thickness.

Although this numerical experiment is simple, it contains the full dynamic coupling between ice dynamics and mass balance field. The intersection of equilibrium line and glacier surface is determined by the ice thickness which depends on the flow. Sliding and deformation vary in a different way with x.

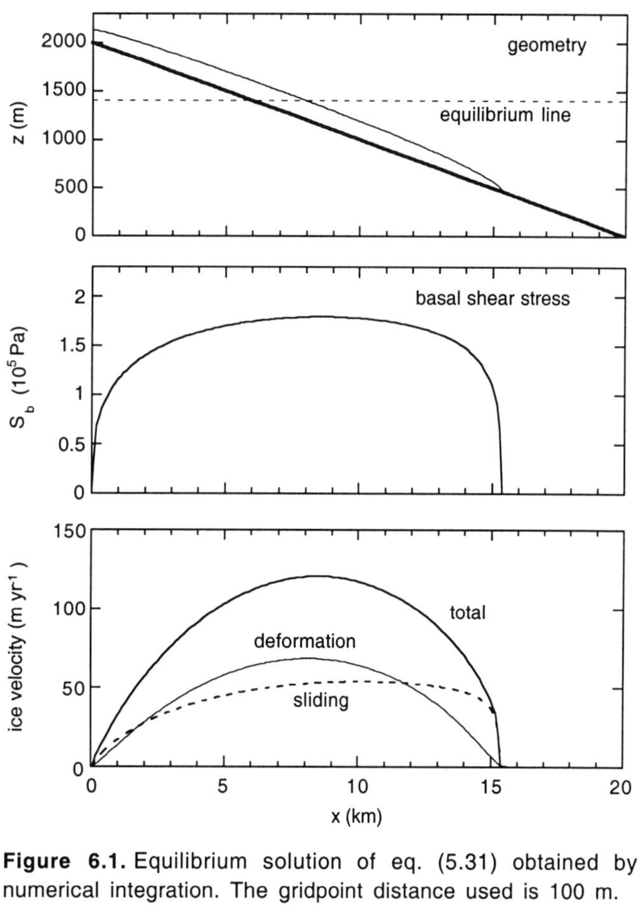

Figure 6.1. Equilibrium solution of eq. (5.31) obtained by numerical integration. The gridpoint distance used is 100 m.

Thus the profile of a glacier with only sliding will be different from that of a glacier with only deformation. *Figure 6.2* shows a comparison of the previous solution with a solution without sliding. This has been obtained by simply setting f_s=0. The difference is small, but significant. However, it is not at all difficult to adjust f_d within its range of uncertainty in such a way that the same glacier length is obtained. The corresponding thickness profiles are shown in *figure 6.3.*

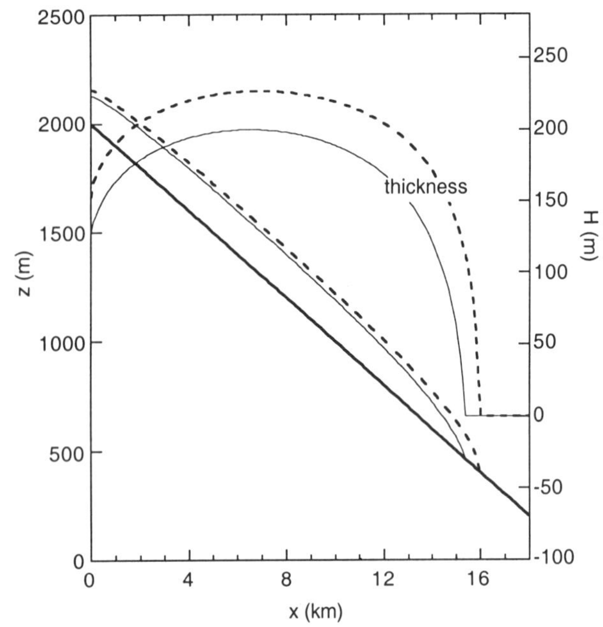

Figure 6.2. A comparison of the full solution with a solution in which no sliding is allowed (dashed). Scale on the right is for ice thickness.

These results make it clear that there is little advantage to be gained from including a sophisticated scheme for sliding in this type of glacier model. There is no need to include longitudinal stresses either. The reason is that the glacier profile is determined basically by the integral of the mass balance along the flow line. It is only in the case of very small bed slopes that changes in the thickness profile associated with different formulations of the flow become significant (for polar ice sheets this is clearly the case).

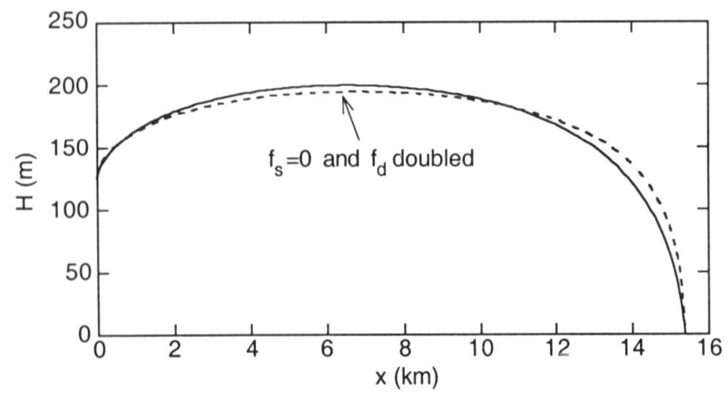

Figure 6.3. Two solutions compared: one without sliding (dashed), one with sliding and deformation.

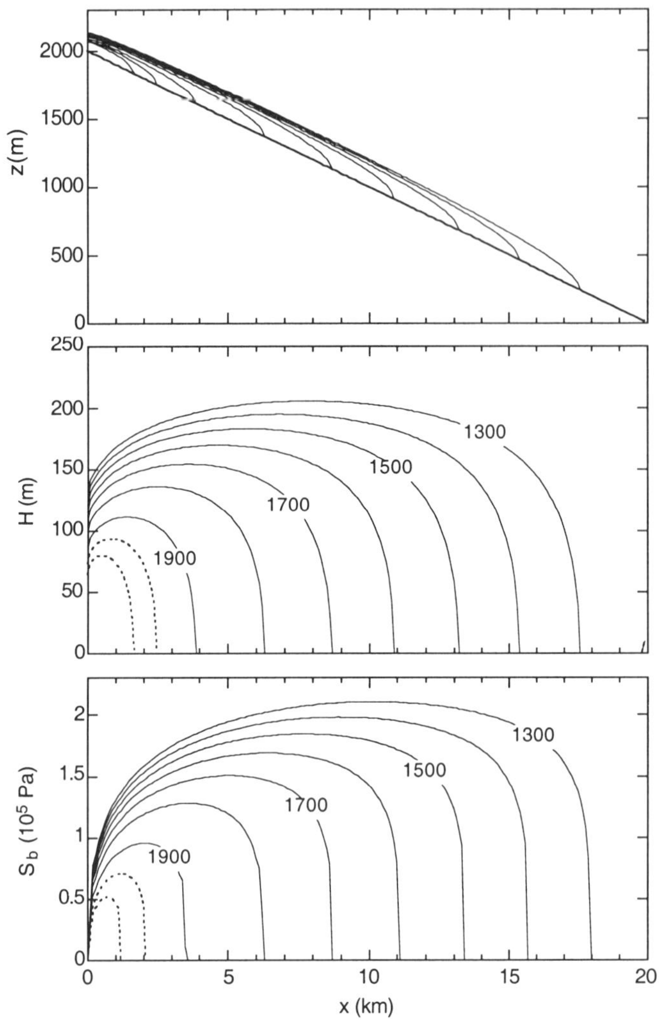

Figure 6.4. A set of solutions for different values of E (labels on the curves). Dashed lines are for E=1975 and E=1950 m.

6.3 Varying the equilibrium-line altitude

Varying the equilibrium-line altitude E is a straightforward matter. Results for a bed slope of 0.1 are shown in *figure 6.4.* The solutions exhibit a strong regularity. It is also interesting to note that the characteristic basal shear stress increases significantly with the size of the glacier, irrespective of the fact that the balance gradient is the same in all calculations.

The analysis can be repeated for different values of the slope of the bed (s). In *figure 6.5* results are shown for s = 0.06, 0.10, 0.14. To facilitate a comparison with the analytical model of section 5.4, L is plotted as a function of b_o - E, where b_0 is the height of the bed at the head of the glacier (x = 0). The dashed lines show the analytical solution with $S_0 / (\rho g) = 10$ m. Clearly, the analytical model underestimates dL/dE, especially for a smaller slope of the bed. A further analysis of the numerical results shows that

$$\frac{dL}{dE} \approx - s^{-3/2} \quad , \qquad (6.2)$$

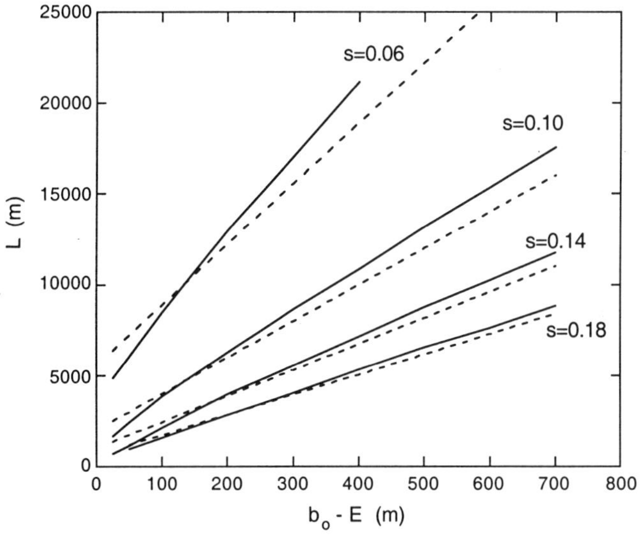

Figure 6.5. Dependence of glacier length on equilibrium-line altitude (b_0 is the highest point of the bed, i.e. at x=0; compare *figure 5.6*). The dashed line shows the solution from the simple analytical model.

whereas for the analytical model we found $dL/dE = 2\ s^{-1}$.
The reason for the greater sensitivity of the numerical model is the dependence of ice thickness on glacier length. In the analytical model the mean ice thickness H_m depends only on the slope of the bed, but in reality it also depends on L.
In *figure 6.6* H_m is plotted versus L. Each data point in the figure represents a steady state for a certain value of E. For instance, the nine data points for s = 0.10 correspond to the profiles shown in *figure 6.4.* Also shown are power law fits. The exponents of these are 0.44 (for s = 0.06), 0.42 (for s = 0.10), 0.41 (for s = 0.14) and 0.40 (for s = 0.18). The relation between glacier length and base stress follows a similar pattern, and therefore it is not shown.

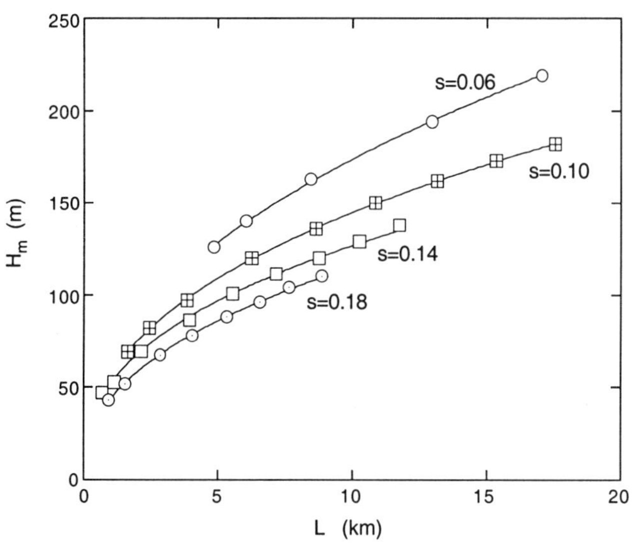

Figure 6.6. The relation between mean glacier thickness and length. Each symbol refers to a model run. The lines show power law fits.

6.4 Back to the analytical model

It is tempting to extend the analytical glacier model of section 5.4 by introducing the dependence of H_m on L. A suitable expression for H_m appears to be

$$H_m = \left(\frac{\mu L}{1+\nu s}\right)^{1/2}, \tag{6.3}$$

where μ and ν are positive constants to be determined by comparison with the results of the numerical model. Eq. (6.3) has been chosen such that for s = 0 it equals eq. (5.18), giving the mean thickness for a perfectly plastic glacier on a flat bed. The exponent 1/2 in eq. (6.3) does not differ too much from the exponents of the power law fits mentioned above.
In the simple analytical model, the expression for L was:

$$L = \frac{2\,(H_m + b_0 - E)}{s} \quad . \tag{6.4}$$

Substituting eq. (6.3) yields (with $E' = E - b_0$):

$$L = \frac{2}{s}\left\{\left(\frac{\mu L}{1+\nu s}\right)^{1/2} - E'\right\} . \tag{6.5}$$

This quadratic equation is most conveniently solved by setting $N = L^{1/2}$. We then have

$$N^2 - \frac{2\,\mu^{1/2}}{s\,(1+\nu s)^{1/2}}\,N + \frac{2E'}{s} = 0 \ . \tag{6.6}$$

The determinant is

$$Det = \frac{4\,\mu}{s^2\,(1+\nu s)} - \frac{8E'}{s} \quad . \tag{6.7}$$

Real solutions exist only when Det $\geq$ 0. Because the first term is always positive, a real solution exists even for small positive values of E', that is, when the equilibrium line is higher than the highest part of the bed at x = 0, but below the glacier surface. This nonlinearity, of course, reflects the feedback between altitude and mass balance.
The solution for L reads

$$L = \left\{\frac{\mu^{1/2}}{s\,(1+\nu s)^{1/2}} \pm \left(\frac{\mu}{s^2\,(1+\nu s)} - \frac{2E'}{s}\right)^{1/2}\right\}^2 \quad . \tag{6.8}$$

Note that values of L for which N<0 are spurious and should not be considered.
To plot the solution, values should be assigned first to μ and ν. The output of the numerical model displayed in *figure 6.6* is represented fairly well by choosing μ = 9 m and ν = 20.
Figure 6.7 shows the resulting L(E')-curve for a bed slope of 0.06. Because L = 0 is a stable solution for E' > 0 (equilibrium line above the bed everywhere), there are two stable branches

and one unstable branch (dotted). In terms of catastrophe theory this model represents a fold, but because we add the condition that L should be positive it appears as a distorted cusp. The branching of the steady-state solutions that shows up here was in fact found numerically by the author in an earlier study [2]. It should also be mentioned that the dynamics of the present glacier model are similar to those of a perfectly plastic ice sheet on a flat bed with a sloping equilibrium line [1].
The range of E'-values for which two stable solutions exist is

$$0 \le E' < E'_{crit} \;, \tag{6.9}$$

where the critical point E'_{crit} is found by setting Det=0:

$$E'_{crit} = \frac{\mu}{2s(1+\nu s)} \;. \tag{6.10}$$

Therefore, for increasing slope of the bed, the height - mass balance feedback becomes weaker and the critical point approaches the origin.
The results obtained with the linear analytical model and the numerical model are also shown. For the linear case $S_0/(\rho g)$ was set to 8 m, because this is consistent with eq. (6.3) for s = 0. Clearly, the nonlinear model performs quite well.
The hysteresis around E' = 0 was not investigated in detail with the numerical model, because this requires another model design with much higher resolution. It seems that in the numerical model the value of E'_{crit} is somewhat smaller. However, we should realise that now we are looking at very small glaciers. The usefulness of the nonlinear model lies in its ability to correctly predict the sensitivity (dL/dE') for a wide range of values of E'. The results for other values of the bed slope are not shown, because the picture is very similar.

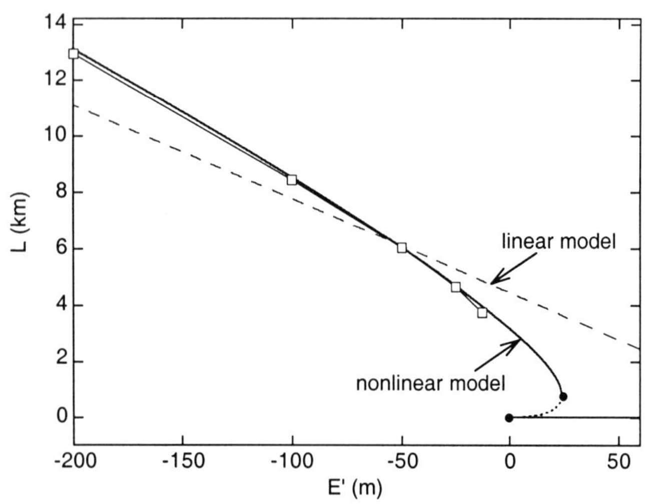

Figure 6.7. The solution of the nonlinear glacier model, eq. (6.5), compared with the linear solution and the results of the numerical model (squares). The bifurcation points are shown by dots.

The climate sensitivity of glacier length is found to be:

$$\frac{dL}{dE} = -\frac{2}{s}\left\{1 + \frac{\Omega}{\left(\Omega^2 - \frac{2}{s}E'\right)^{1/2}}\right\} , \tag{6.11}$$

where

$$\Omega = \frac{\mu^{1/2}}{s(1+\nu s)^{1/2}} \tag{6.12}$$

Thus dL/dE now depends on E', which was not the case with the linear model. In fact, the expression between { } represents the nonlinear correction to the linear result (-2/s); eq. (5.23)). For increasing (negative) values of E' and/or increasing slope of the bed the result approaches the linear case. The result is illustrated in *figure 6.8*. In conclusion we can state that the most sensitive glaciers are small glaciers on a bed with a small slope.

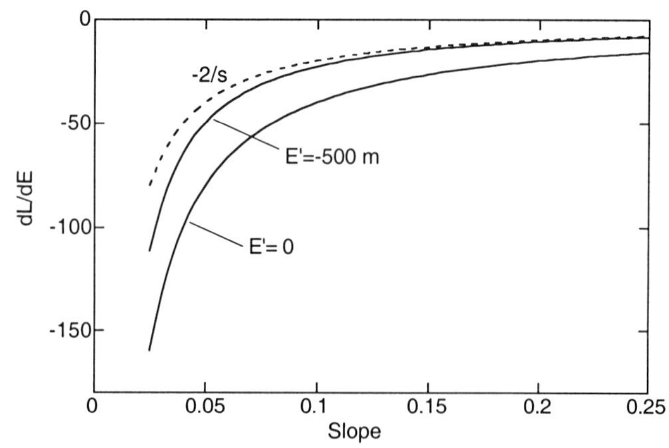

Figure 6.8. The sensitivity of glacier length to changes in the equilibrium-line altitude. Solid lines: nonlinear model; dashed line: linear model.

6.5 A more complicated bed

In the previous sections we considered glaciers on a bed with a constant slope. In general this is not a good approximation to the real situation, of course. Before real bed profiles are considered in Chapter 8, it is instructive to do some calculations with the numerical model for a slightly more complicated bed. A possibility is to consider a piecewise linear slope, as shown in *figure 6.9*: the glacier bed is steeper in the upper part (s = 0.15) and gentle (s = 0.03) in the lower part. This situation is not uncommon when we look at the shape of larger valley glaciers during the little ice-age advance.
On the basis of the foregoing analysis, we expect that a glacier on such a bed will become more sensitive to changes in E when a significant part of the gentle slope is covered by ice. This is indeed what happens. The three profiles shown in *figure 6.9* illustrate the asymmetry. Using the profile for E=1500 m

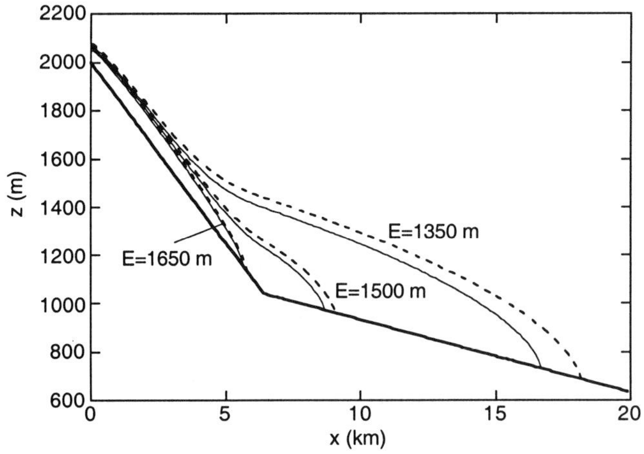

Figure 6.9. Steady-state profiles of glaciers on a bed with a kink, for different values of the equilibrium-line altitude E. Solid lines for β=0.007 a^{-1} dashed lines for β=0.01 a^{-1}.

as a reference, a 150 m increase in E makes the glacier about 2.5 km shorter, whereas a 150 m decrease in E makes it 8 km longer!

Figure 6.10 shows steady-state glacier length as a continuous function of E. The change in sensitivity is gradual. For a high equilibrium line dL/dE is typically 15, but it increases to 50 when the equilibrium line drops to 1400 m. Therefore there is a threefold increase in the sensitivity of this glacier when the climate gets colder. Glacier volume is also shown, normalised in such a way that it is equal to glacier length for E = 1325 m. As expected, the variation in volume is larger than the variation in length. It should be noted that, if the slope of the lower section were zero, the glacier would grow to infinity once it passed a critical point! This example makes it clear that the shape of the bed has a profound influence on the equilibrium states of a glacier and on its sensitivity to climate change. So in studies of real glaciers it is essential to parameterise and incorporate the real geometry in the model.

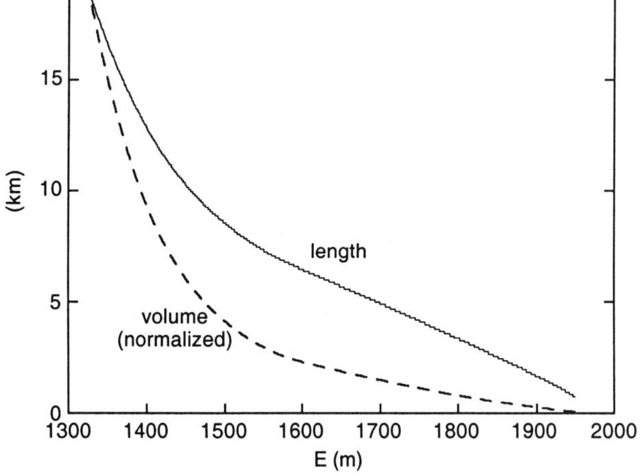

Figure 6.10. Steady-state glacier length and volume (dashed) for the bed profile shown in *figure 6.9*. (β=0.007 a^{-1}).

6.6 Characteristic time scales

So far we have considered only equilibrium states. Most of the time interest is in the transient behaviour of glaciers, because the forcing is constantly changing. The numerical model described in Chapter 5 is a useful tool for investigating how glaciers respond to changes in mass balance.

In the glaciological literature there is no real consensus about the definition of response. Reaction time, response time, relaxation time, etc. are mentioned, but not defined in a standardised way. In fact, different studies use different definitions and this book is no exception.

Let V_i be the equilibrium volume of a glacier, with equilibrium length L_i, that is in perfect balance with the prevailing (constant) climatic state C_i. We then define characteristic time scales as follows:

Growth time t_g**:**
The growth time is the time a glacier in a constant climatic state C_i needs to attain a volume of $(1-e^{-1})V_i$, starting from zero ice volume.

Relaxation time t_{rel}:
Consider a glacier in a constant climatic state where a small part of the mass δV is removed. The relaxation time is the time a glacier needs to return to equilibrium, i.e. to attain a volume $V_i - e^{-1}\,\delta V$.

Response time t_{rV}:
The climatic state is changed stepwise from C_1 to C_2. The corresponding equilibrium glacier volumes are V_1 and V_2. The response time is the time a glacier needs to attain a volume $V_2 - e^{-1}(V_2-V_1)$.

Length response time t_{rL}**:**
As above for glacier length L.

The various time scales may differ significantly. The growth time is particularly long when bed slopes are very small, because of the height - mass balance feedback. At the same time, the relaxation time can be small. For most glaciers relaxation and response times have the same order-of-magnitude. Glacier length and volume have slightly different response times, the volume response time being smaller.

The characteristic time scales defined above are e-folding time scales, based on the general concept that a relaxation time is defined for a system linearised around a given equilibrium state. In view of this, perturbations introduced to calculate the time scales with a numerical model should be sufficiently small. Here sufficiently means that the calculated time scales become independent of the magnitude of the perturbation.

Can t_{rV} and t_{rL} be estimated for a particular glacier without detailed numerical modelling? Early attempts to infer values of t_{rV} from theoretical considerations were based on linear kinematic wave theory [3]. Details of this theory are not given here. The most important result of this theory is that t_{rV} is of the order of L/U_{sn}, where L is glacier length and U_{sn} ice velocity at the snout. This yields characteristic response times of 100 to 1000 years for valley glaciers. Several investigators have questioned the validity of this approach. A problem is that U_{sn} is poorly defined, or at least difficult to estimate for real glaciers. There is also substantial field evidence from valley glaciers which suggests response times between 10 and 50 years rather than the large values predicted by the linear kinematic wave theory [4].

Jóhannesson and co-workers [5] have provided an alternative analysis of the problem. They make it plausible that a good estimate of the response time (which they refer to as the volume time scale) can be obtained from the expression:

$$t_{rV} = \frac{H^*}{-B_{term}} \quad . \tag{6.13}$$

Here H* is a characteristic ice thickness and B_{term} the specific balance at the terminus. They also show that, for regular valley glaciers, the details of the snout dynamics are not important for the global behaviour of the glacier and provide theoretical support for what many glaciologists have assumed intuitively.

Numerical models of the type described in Chapter 5 have yielded values of t_{rV} or t_{rL} in line with observations [6], inspiring some confidence in the dynamics described by eq. (5.31). We will therefore explore the model with the kinked bed profile a bit further and calculate the various time scales for the three glaciers shown in *figure 6.9*. The results are listed in *table 6.1*. The relaxation time was computed by removing instantaneously 5 m of ice over the entire length of the glacier. A 50 m change in E appeared to be sufficiently small to obtain the response times.

Clearly the growth time is longer for the larger equilibrium states, but is not in proportion to the steady-state volumes. The response times for volume and length show a consistent pattern: t_{rL} is 1.35 times t_{rV} for all three cases.

Finally, we should realise that the particular behaviour of the glacier snout tells us little about response time. When climatic cooling sets in, a glacier front will sooner or later start to advance. The time difference between climatic deterioration and the start of the advance has sometimes been called reaction time. However, because this time difference depends on the past evolution of the glacier, it certainly does not reflect a well-defined physical property of the glacier.

E (m)	L (km)	t_g	t_{rel}	t_{rV}	t_{rL}
1650	5.55	99	53	39	71
1500	8.35	155	122	68	89
1350	16.35	408	167	139	201

Table 6.1. Characteristic time scales for the glaciers shown in *figure 6.9*. Volume is normalised as in *figure 6.10*. All time scales are in years (a).

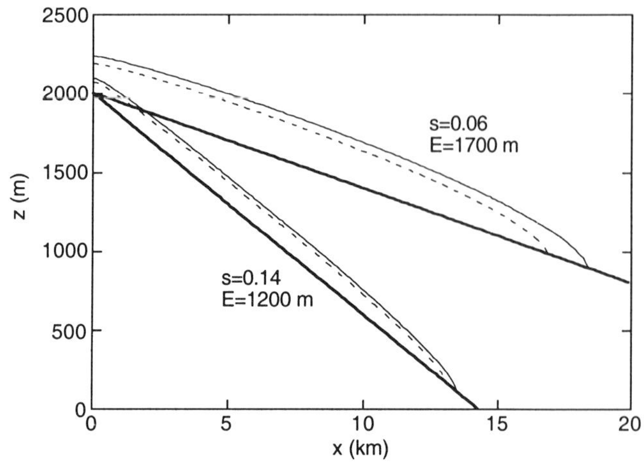

Figure 6.11. Effect of the balance gradient on the equilibrium profile, shown for two different slopes of the bed (0.14 and 0.06). The continuous line corresponds to a balance gradient of 0.01, the dashed line to a balance gradient of 0.005.

6.7 The role of the balance gradient

In the calculations presented so far, the balance gradient had a constant value of 0.007 a^{-1}. Although this can be considered as a characteristic value, see *figure 4.4*, observed values range from 0.002 to 0.01 a^{-1}. Additional calculations have therefore been made to find the effect of the balance gradient. In *figure 6.9*, equilibrium profiles are shown for a = 0.01 and a = 0.007 a^{-1}. The larger balance gradient leads to longer glaciers. The same result is found for glaciers on beds with a constant slope (*figure 6.11*). Apparently, the fact that glaciers become larger when the balance gradient increases is a general result. This must operate through the height-mass balance feedback, so glaciers tend to be thicker when the balance gradient is larger. The effect is not very strong, however.

Concerning the mass flux it is a different story, of course. Changing the balance gradient from 0.01 to 0.005 a^{-1} reduces the mass flux for any x by about 50%. Clearly, this must be so to conserve mass. Because ice velocity depends so strongly on ice thickness, see eq. (5.29), only a small reduction in ice thickness is needed to achieve the decrease of the flux. This explains why equilibrium length and mean ice thickness depend only weakly on the balance gradient.

The influence of the balance gradient on the time scales appears to be quite significant. Here only one result is shown, namely the response time for the larger glacier in *figure 6.11*. *Figure 6.12* shows the response of glacier length to a sudden 100 m increase in the equilibrium-line altitude, yielding t_{rL} = 74 a. Repeating the experiment with a balance gradient of 0.005 a^{-1} yields t_{rL} = 134 a. To make a meaningful comparison, E was first set to 1665 m (instead of 1700 m). So, with regard to the equilibrium state, a 35 m change in E is needed to compensate for the (strongly) reduced balance gradient.

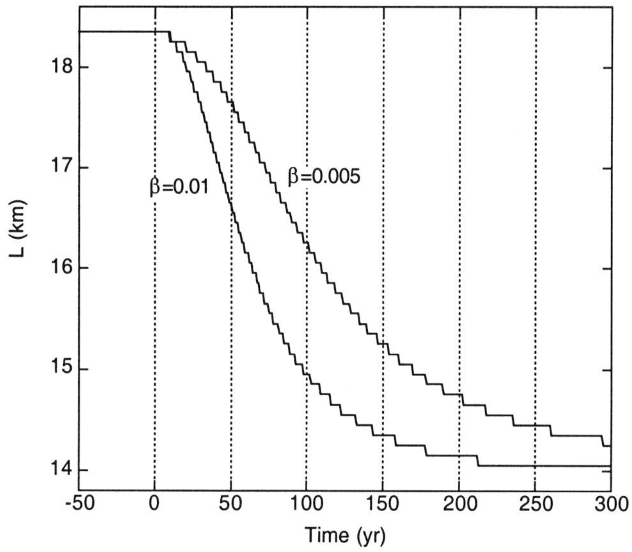

Figure 6.12. The response of a glacier on a bed with $s = 0.06$ to a sudden 100 m increase in E at time zero. Glacier length is shown for two different values of the balance gradient β.

This result and other results (not shown here) make it clear that the response time increases significantly when the balance gradient is smaller.

6.8 Response time derived from the simple analytical model

From a general point of view the adjustment of the glacier length to a change in the equilibrium-line altitude requires a decrease or increase in the mass transfer down-glacier. The speed at which this can be accomplished is determined by the characteristic ice velocity $\tilde{u}$. The adjustment has to find its way down to the glacier front, so L seems to be an appropriate length scale. Thus we may estimate t_{rL} in the following way:

$$t_{rL} \approx c \frac{L}{\tilde{u}} \, , \qquad (6.14)$$

where c is a dimensionless constant.
It is not difficult to work out the implications of eq. (6.14) for the analytical glacier model. With reference to *figure 5.6* the volume flux at the equilibrium line is

$$H_E U_E = \beta \int_0^{L/2} (H - s x - E')\, dx \; . \qquad (6.15)$$

Here H_E is the ice thickness at the equilibrium line. Evaluating the integral yields

$$H_E U_E = \beta \int_0^{L/2} (H - s x - E')\, dx = \frac{\beta}{2} \left\{ (H_m - E')\, L + \frac{s}{4} L^2 \right\} \; . \qquad (6.16)$$

H_m is the mean thickness of the glacier in the accumulation zone. If H_m is assumed to be identical to the overall mean ice thickness the condition for equilibrium reads (eq. (5.21)):

$$H_m - E' = \frac{s L}{2} \; . \qquad (6.17)$$

Combining eqs. (6.16)-(6.17) gives

$$U_E = \frac{3 \beta s L^2}{8 H_E} \; . \qquad (6.18)$$

We then find for the response time

$$t_{rL} = \frac{c L}{U_E} = \frac{8 c H_E}{3 \beta s L} \; . \qquad (6.19)$$

This expression still contains ice thickness at the equilibrium line (H_E). For a glacier on a bed with a constant slope it is quite reasonable to assume that H_E is proportional to H_m, so

$$H_E = c_1 H_m = c_1 \left(\frac{\mu L}{1 + \nu s} \right)^{1/2} . \qquad (6.20)$$

Substituting eq. (6.20) into eq. (6.19) then yields

$$t_{rL} = \frac{8 c_2 \mu^{1/2}}{3 \beta s \, (1 + \nu s)^{1/2} L^{1/2}} \, , \qquad (6.21)$$

where we have written $c_2 = c\, c_1$ This expression relates the response time to glacier length, bed slope and balance gradient. The constant c_2 should be of order 1.
At first sight eq. (6.21) seems curious because it predicts decreasing response time for increasing glacier length. In fact, such a result is also found from rigorous scaling of the dynamic equations governing glacier flow [7].
We should realise, however, that long glaciers are generally on beds with small slopes! As a result the denominator of eq. (6.21) is more constant than one would expect at first sight. We can investigate this further by relating L to s for a given value of E. For the nonlinear model we should make use of eq. (6.5). However, s cannot be solved from this equation analytically, so we prefer to use eq. (5.21) from the linear model and write

$$s = \zeta L^{-1} , \text{ where } \zeta = 2\,(H_m + b_0 - E) \; . \qquad (6.22)$$

The response time then becomes

$$t_{rL} = \frac{8 c_2 \mu^{1/2} L^{1/2}}{3 \beta \zeta \left(1 + \nu \zeta L^{-1}\right)^{1/2}} \; . \qquad (6.23)$$

Figure 6.13 shows t_{rL} for three values of ζ. The parameters μ and ν have been given the same values as before ($\mu = 9$ m,

$\mu = 30$). The balance gradient was set to 0.00777 m ice a^{-1} (corresponding to 0.007 mwe a^{-1}) and $c_2 = 1$. Apparently, now the response time increases with glacier length!

Larger glaciers not only have smaller slopes, they also tend to be in regions with the highest mountains. Since ζ is essentially twice the altitudinal difference between the head of the glacier and the equilibrium line, ζ will tend to have larger values for longer glaciers. The effect would be a somewhat weaker dependence of t_{rL} on L than depicted in *figure 6.13*.

Eq. (6.21) also shows that the response time is inversely proportional to the balance gradient, which is in qualitative agreement with the results of the numerical model.

The issue of response times will be discussed again in Chapter 8, when the modelling of real glaciers is discussed.

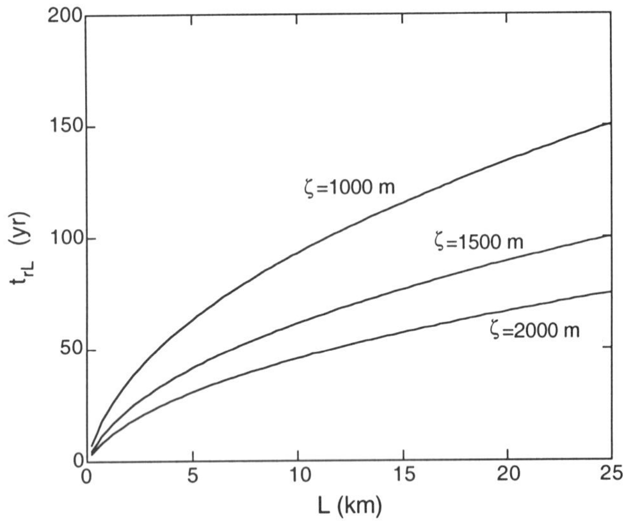

Figure 6.13. Response time for glacier length calculated from eq. (6.23).

References and further reading

1. J. Weertman (1961): Stability of ice-age ice sheets. *Journal of Geophysical Research* **66**, 3783-3792.

2. J. Oerlemans (1981): Some basic experiments with a vertically-integrated ice-sheet model. *Tellus* **33**, 1-11.

3. J.F. Nye (1960): The response of glaciers and ice-sheets to seasonal and climatic changes. *Proceedings of the Royal Society of London* **A256**, 559-584.

   J.F. Nye (1965): The frequency response of glaciers. *Journal of Glaciology* **5**, 567-587.

4. R.S.W. van de Wal and J. Oerlemans (1995): Response of valley glaciers to climate change and kinematic waves: a study with a numerical ice-flow model. *Journal of Glaciology* **9**, 115-131.

5. T. Jóhannesson, C.F. Raymond and E.D. Waddington (1989): Time-scale for adjustment of glaciers to changes in mass balance. *Journal of Glaciology* **35**, 355-369.

6. W.F. Budd and D. Jenssen (1975): Numerical modelling of glacier systems IAHS-AISH Publ. No. 104, 257-291.

   P.D. Kruss (1983): Climatic change in East Africa: numerical modeling from the 100 years of terminus record of Lewis Glacier, Mount Kenya. *Zeitschrift für Gletscherkunde und Glazialgeologie* **19**, 43-60.

   J. Oerlemans (1986): An attempt to simulate historical front variations of Nigardsbreen, Norway. *Theoretical and Applied Climatology* **37**, 126-135.

7. D.B. Bahr and M. Dyurgerov (1999): Characteristic mass-balance scaling with valley glacier size. *Journal of Glaciology* **45** (149), 17-21.

7. More realistic geometry

7.1 Area-elevation distribution

So far we have only modelled glaciers of constant width, because we need a basic understanding of simple situations before being confronted with more complicated geometry. We will now relax this constraint and consider calculations in which the width of a glacier is allowed to vary. This requires an adjustment of the continuity equation, to be discussed shortly. First we look at the area-elevation distribution of some typical glaciers.

Figure 7.1 (next page) gives an impression of the variety seen in reality. The plots have been normalised in such a way that they all have the same area. The equilibrium line is indicated by a horizontal line. In a way, two typical types of distribution can be identified in this sample. The first type is highly asymmetric with respect to the equilibrium line (Nigardsbreen, Engabreen and Franz-Josef Glacier). Ice drains from a wide accumulation basin into a relatively narrow valley. The equilibrium line is at a significantly lower altitude than the mean elevation of the glacier. This reflects the fact that the mean ablation rate is much higher than the mean accumulation rate, in spite of the fact that Nigardsbreen, Engabreen and Franz-Josef Glacier are maritime glaciers with high precipitation. The second type is more symmetric with respect to the equilibrium line. From there the area falls off both down- and up-glacier.

The area-elevation distribution is not necessarily an exact measure of how glacier width varies. It would only be so if the surface slope were constant along the entire flow line. Nevertheless, distributions of the type shown in *figure 7.1* reflect the fact that, in general, accumulation areas are wider than ablation areas. In a way one can argue that this is implied in the definition of a valley glacier. But even for ice caps that terminate mainly on land, like Devon Ice Cap (Canadian Arctic), Vatnajökull (Iceland) and Hardangerjøkulen (Norway) the area-elevation distribution is similar.

To what extent do the dynamics of these glaciers resemble the dynamics of a glacier of constant width? The answer has to come from explicit numerical modelling. Some insight can be gained, however, by a modest extension of the simple analytical glacier model discussed in section 5.4.

7.2 Extending the simple analytical model

Figure 7.2 illustrates the idea. A glacier tongue of constant width is nourished by an upper basin of length L_{ub}, also of constant, but different, width. To find the general solution we have to consider two cases, namely $L \leq L_{ub}$ and $L > L_{ub}$.

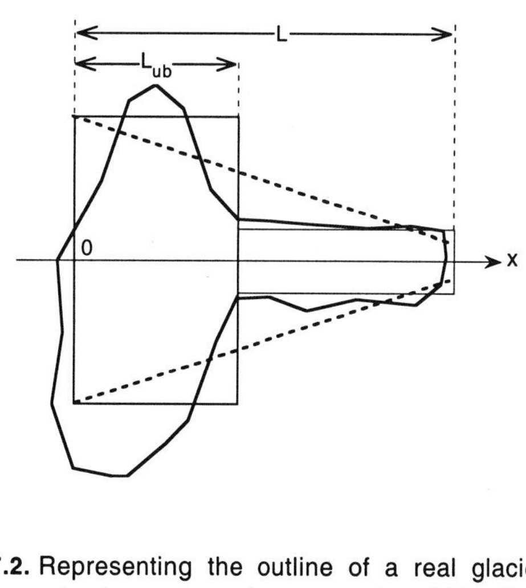

Figure 7.2. Representing the outline of a real glacier (solid line) by a simple rectangular geometry (thin line). The dashed line shown an alternative representation with steadily decreasing width.

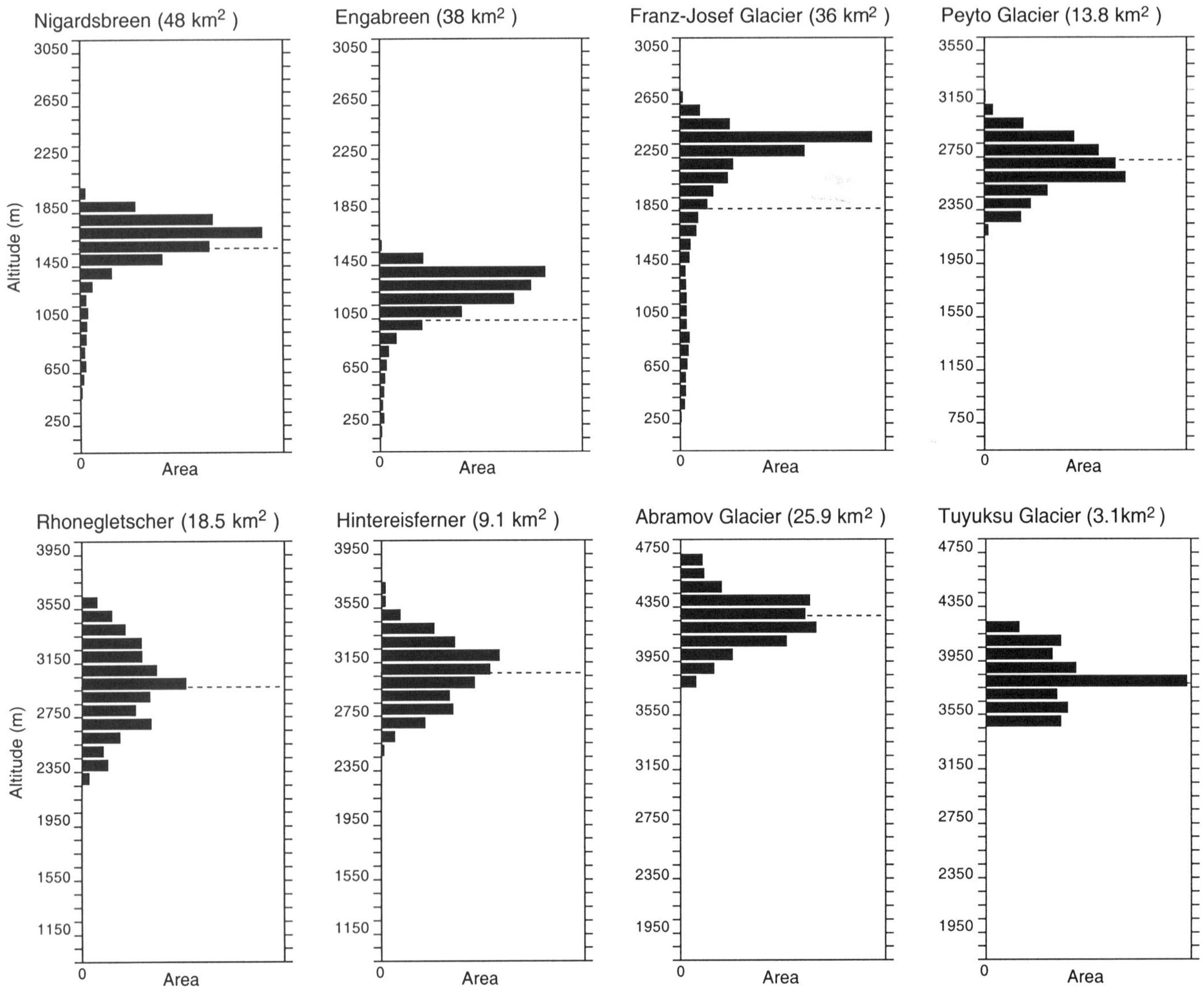

Figure 7.1. Area-elevation distribution for a number of glaciers. In each case the horizontal scale has been adjusted to give the same area in the diagram. The dashed line shows the approximate equilibrium-line altitude. Data from the World Glacier Monitoring Service (Zürich) with some additions.

For $L \leq L_{ub}$ the solution is given by eq. (5.21):

$$L = \frac{2\,(H + b_0 - E)}{s} \quad . \tag{7.1}$$

As before, H is ice thickness, s is slope of the bed, b_0 is bed elevation at the head of the glacier, and E is the equilibrium-line altitude.

When the mass budget of the upper basin is positive L will be larger than L_{ub}. The length of the glacier is then determined by

$$\int_0^{L_{ub}} (H + b_0 - s\,x - E)\,dx \; + \xi \int_{L_{ub}}^{L} (H + b_0 - s\,x - E)\,dx = 0 \quad . \tag{7.2}$$

In this expression ξ is the width of the glacier tongue scaled with the width of the upper basin. For most glaciers $\xi < 1$. However, we will also consider the case of a widening glacier ($\xi > 1$). Evaluating the integrals yields:

$$-\tfrac{1}{2}\xi\, s\, L^2 + \xi\,(H + b_0 - E)\,L + (1 - \xi)\,(H + b_0 - E)\,L_{ub} - \tfrac{1}{2} s\,(1 - \xi)\,L_{ub}^2 = 0 \quad . \tag{7.3}$$

With $E' = E - b_0 - H$ $(E' < 0)$, the solution is:

$$L = -\frac{1}{s}\left[E' + \left\{E'^2 + 2\,s\,\frac{1-\xi}{\xi}\left(E'\,L_{ub} + \tfrac{1}{2}\,s\,L_{ub}^2\right)\right\}^{1/2}\right] \quad . \tag{7.4}$$

An example is shown in *figure 7.3.* The upper basin has a length of 5 km. Other parameter values are: $b_0 = 2000$ m, $s = 0.1$, $H = 100$ m. L is plotted as a function of E for three different values of ξ. For $\xi = 0.25$, the width of the upper basin is four times that of the glacier tongue. This is not an unusual situation as illustrated by the area-elevation distribution of Nigardsbreen and Franz-Josef Glacier (see also *figure 5.1*).

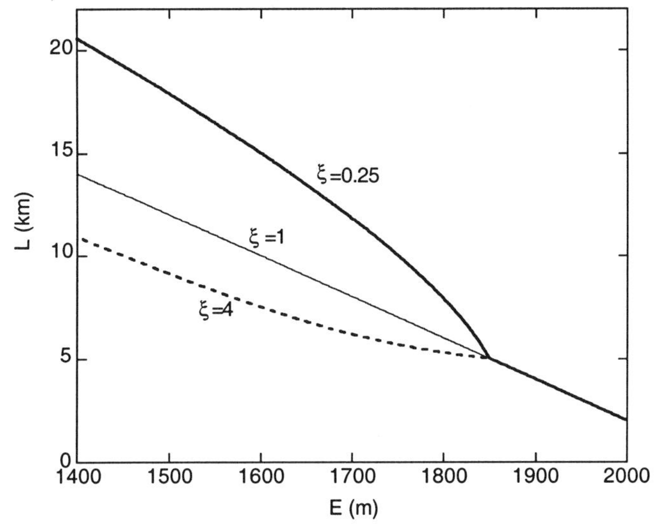

Figure 7.3. Glacier length as a function of the equilibrium-line altitude, depending on the geometry as prescribed by ξ.

Clearly, in this case the glacier length increases rapidly when the equilibrium line sinks below 1850 m. In terms of the climate sensititivy dL/dE, a clear maximum exists when the glacier starts to form the tongue. When the equilibrium line sinks still further, dL/dE will approach the value of the original model glacier of uniform width ($\xi = 1$).

When the lower part of the glacier is wider than the upper part the opposite is seen. The sensitivity is at a minimum when the budget of the upper basin is just positive.

In the foregoing analysis we did not allow the ice thickness to vary with the length of the glacier, as in eq. (6.3) where $H \propto L^{1/2}$. Unfortunately, combining such an expression with eq. (7.3) yields an equation that cannot easily be treated analytically.

In conclusion we can state that glaciers with a narrow tongue are the ones that are most sensitive to climate change, especially when the glacier front is just below the upper basin.

Another geometry that can be handled easily is when the width of a glacier increases or decreases steadily, as in the case of the glacier depicted by the dashed lines in *figure 7.2*. If the width of the glacier is written as $w_0 - p\,x$, where w_0 is the width at the glacier head and p a constant, the condition for equilibrium reads

$$\int_0^L (w_0 - p\,x)\,(H + b_0 - s\,x - E)\,dx = 0 \ . \qquad (7.5)$$

This leads to the equation

$$\tfrac{1}{3} p\, s\, L^2 + \tfrac{1}{2}(p\,E' - s\,w_0)\,L - E'\,w_0 = 0 \ . \qquad (7.6)$$

The solution is

$$L = \frac{-\tfrac{1}{2}(p\,E' - s\,w_0) + \left\{ \tfrac{1}{4}(p\,E' - s\,w_0)^2 + \tfrac{4}{3} p\,s\,w_0\,E' \right\}^{1/2}}{\tfrac{2}{3} p\,s} \ . \qquad (7.7)$$

In *figure 7.4* the solution is plotted for $b_0 = 2000$ m, $s = 0.1$, $w_0 = 4000$ m, $p = 0.2$, $H = 100$ m. The case of uniform glacier width ($p = 0$) is shown for reference. Clearly, the glacier is more sensitive (|dL/dE| larger) when the width of the bed decreases with x ($p > 0$). The solution for $p = -0.2$ is also shown. In this case the sensitivity is slightly reduced.

The models for a simple geometry discussed above can be combined. In fact, almost any valley glacier with a smooth geometry can be handled if we split it up into a number of segments that have constant width or a width that is decreasing or increasing at a constant rate.

A comment needs to be made concerning the ice thickness H. As noted above, an analytical solution cannot be obtained if we make H dependent on L. However, it is not difficult to find the solution by means of an iterative method. Whether a simple relation of the type $H \propto L^{1/2}$ still holds for more complicated geometries is an open question.

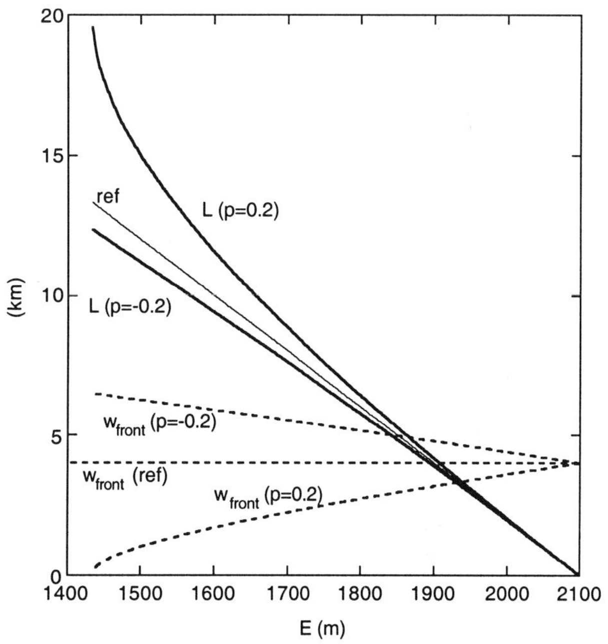

Figure 7.4. Glacier length (L) and width at the glacier front (w_{front}) for the model with a steadily decreasing / increasing width of the bed. The curve labelled 'ref' refers to a glacier of constant width.

7.3 Adjusting the numerical model for varying glacier width

In order to include a three-dimensional geometry in a numerical flow-line model the continuity equation has to be adjusted. Eq. (2.5) can be integrated vertically and perpendicular to the flow line. This yields

$$\frac{\partial A}{\partial t} = -\frac{\partial (AU)}{\partial x} + B \ , \qquad (7.8)$$

where A is the area of the cross section perpendicular to the flow line. Note that eq. (7.8) is only valid when ice density is taken to be constant and curvature effects can be neglected.
Now a shape for the cross profile has to be chosen: a possible shape is trapezoidal (see section 5.1), for which (see *figure 7.5*)

$$A = H\left(w_0 + \tfrac{1}{2}\lambda H\right) \ . \qquad (7.9)$$

Substituting eq. (7.9) into eq. (7.8) then yields a prognostic equation for ice thickness:

$$\frac{\partial H}{\partial t} = \frac{-1}{w_0 + \lambda H}\frac{\partial}{\partial x}\left[(w_0 + \tfrac{1}{2}\lambda H)\,U\,H\right] + B \ . \qquad (7.10)$$

When $\lambda = 0$ and w_0 is constant this reduces to eq. (5.30). Now the simplest way to proceed is to assume that the relation between glacier geometry and vertical mean ice velocity U is the same as for a glacier with lateral symmetry, i.e.

$$U = U_d + U_s = f_d\,H\,S_e^{\,3} + \frac{f_s\,S_e^{\,3}}{H} \ . \qquad (7.11)$$

S_d, the driving stress, has been replaced by an effective driving stress S_e which should deal with the fact that the glacier is not infinitely wide but also experiences drag from the sides. In the next section we will discuss how S_e can be related to S_b.
In summary, the model for a glacier bed of varying width and varying steepness of the valley slopes can be described by the following rate equation for ice thickness:

$$\frac{\partial H}{\partial t} = \frac{-1}{w_0 + \lambda H}\frac{\partial}{\partial x}\left[D\frac{\partial(b+H)}{\partial x}\right] + B \ , \qquad (7.12)$$

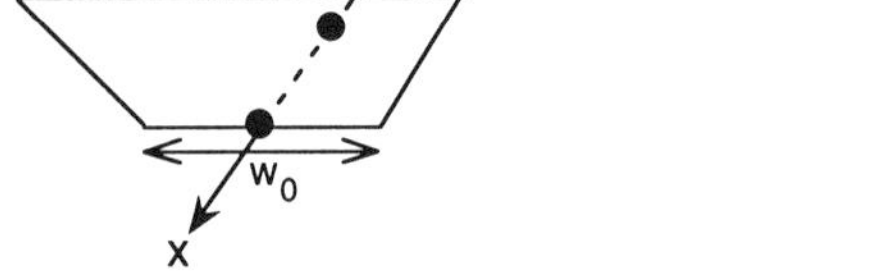

Figure 7.5. A possible parameterisation of the three-dimensional geometry for a flow-line model. The parameters λ and w_0 are prescribed as a function of x (distance along the flow line). Gridpoints are indicated by black dots.

where the diffusivity

$$D = (w_0 + \tfrac{1}{2}\lambda H)\,(\rho\,g)^3\,H^3\left(\frac{\partial h}{\partial x}\right)^2\left(f_d\,H^2 + f_s\right) \ . \qquad (7.13)$$

So the modifications to the model discussed in section 5 are minor, and the numerics can be done in the same way. As for the longitudinal bed profile, it is advisable to perform some averaging in such a way that w_o and λ vary smoothly with x. Too large gradients in these quantities will create large truncation errors in the numerical scheme and may lead to instability or poor conservation of mass in the model.

7.4 Shape factors

The first theoretical treatment of ice flow in an inclined cylindrical channel of infinite length and prescribed cross-sections was given by Nye [1]. He discussed the simplest possible set of equations describing the flow and solved these numerically. A brief summary of his theory is given here, because it provides a way to parameterise the effect of side drag in a model that uses only ice thickness and surface slope at the centre line to compute the driving force.

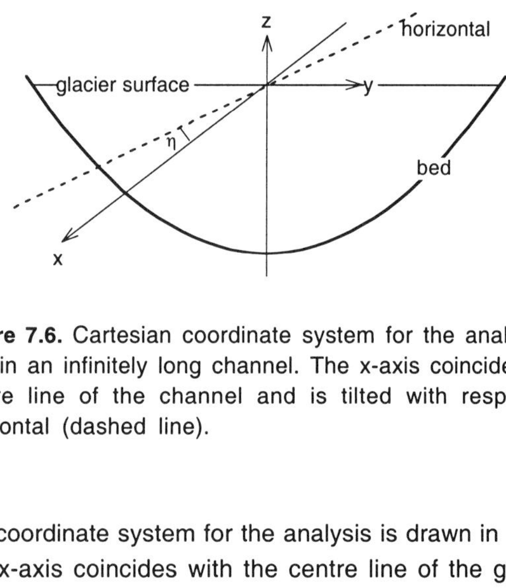

Figure 7.6. Cartesian coordinate system for the analysis of ice flow in an infinitely long channel. The x-axis coincides with the centre line of the channel and is tilted with respect to the horizontal (dashed line).

The coordinate system for the analysis is drawn in *figure 7.6*. The x-axis coincides with the centre line of the glacier and makes an angle α with the horizontal. A further assumption is that the velocity u is in the x-direction everywhere. The stress and velocity fields do not vary with x, so the flow can still be classified as simple shear.
In the 1-dimensional analysis of section 5.2 the only nonzero stress component was S_{xz}. With the assumptions mentioned above there are two stress components: S_{xz} and S_{xy}, and the equilibrium of forces becomes

$$\frac{\partial S_{xy}}{\partial y} + \frac{\partial S_{xz}}{\partial z} + \rho\,g\sin\eta = 0 \ . \qquad (7.14)$$

Glen's law for plain shear now takes the form

$$\frac{\partial u}{\partial y} = A\left(S_{xy}^2 + S_{xz}^2\right)^n S_{xy} \ , \tag{7.15}$$

$$\frac{\partial u}{\partial z} = A\left(S_{xy}^2 + S_{xz}^2\right)^n S_{xz} \ . \tag{7.16}$$

Eqs. (7.14)-(7.16) have to be solved for S_{xy}, S_{xz} and u, for appropriate boundary conditions. These are u = 0 at the glacier bed and $S_{xz} = 0$ at the glacier surface. For a semi-circular channel of radius R the solution is easily derived. It reads

$$u = \frac{AR}{n+1} \{\rho g \sin\eta\}^n \left[1 - \left(\frac{r}{R}\right)^{n+1}\right] , \tag{7.17}$$

where $r = (y^2 + z^2)^{1/2}$.
Apparently, in this case the velocity profile is axi-symmetric and varies with r as shown in *figure 5.4* (note that in the analysis of section 5.2 the origin was at the glacier bed - not at the surface).
For more complicated geometries numerical methods have to be used. Nye designed a scheme that appeared to work well. We will not discuss his scheme, but will look only at some of his results. In the present context it is most relevant to look at the shape factor F, defined as

$$S_e = F \rho g H \sin\alpha = F S_b \ . \tag{7.18}$$

Nye did the calculations for a rectangular, parabolic and semi-elliptic cross-section, and some results are summarised in *figure 7.7*. F is plotted as a function of W/H, where H is ice thickness at y = 0 and W is the width of the glacier. For increasing W/H, the shape factor approaches 1 rather rapidly. The factors do not vary much between the different shapes. In fact, these results suggest that a general parameterisation for use in numerical flow-line models makes sense. A possible formulation is

$$F = \frac{2}{\pi} \operatorname{arctg}\left(\frac{W}{2H}\right) \ . \tag{7.19}$$

This curve is shown in *figure 7.7* as well.
In this analysis there is no sliding. Although difficult to prove in a rigorous way, it is plausible that similar results emerge when sliding is included. Nye also notes that, at least for a parabolic cross-section, the mean velocity at the surface (averaged from y = 0 to y = W/2) is very close to the velocity averaged over the entire cross-section. The difference is not larger than a few percentage points. This result is independent of the relative contribution made by sliding to the flow.
In the following section we will investigate to what extent the results obtained with the numerical model will change if a shape factor as defined in eq. (7.19) is included.
A more comprehensive discussion on lateral drag and shape factors can be found in Van der Veen [2].

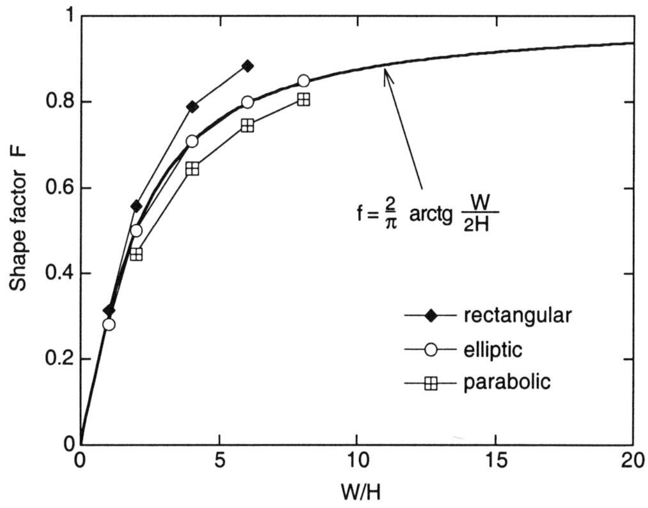

Figure 7.7. The shape factor as calculated by Nye for three different cross-sections [1]. A general fit is also shown.

7.5 Numerical experiments for varying glacier width

It is interesting to repeat the calculation for the steadily decreasing glacier width with the numerical model. *Figure 7.8* shows how the equilibrium glacier length depends on E. The model parameters are $b_0 = 2000$ m, s = 0.1, $w_0 = 4000$ m and $\beta = 0.07\ a^{-1}$. The solution is quite similar to that obtained with the analytical model (compare with *figure 7.4*). For p = 0.2 the sensitivity of the glacier increases when the tongue becomes narrower. For the reference case (p = 0), L varies almost linearly with E, as found earlier.

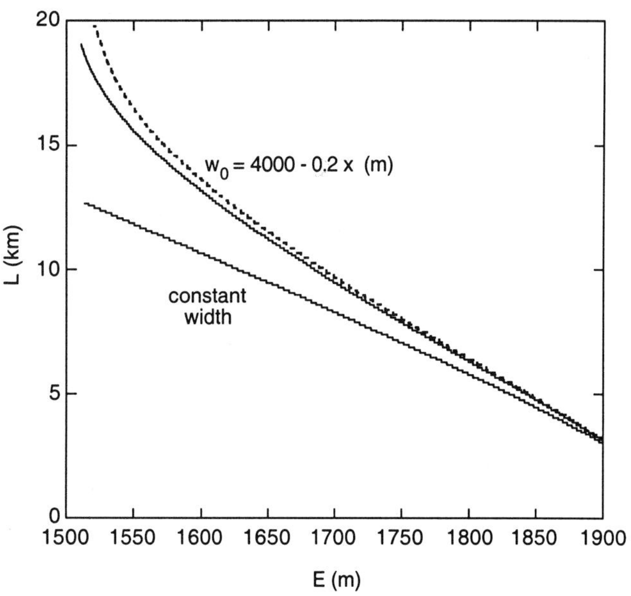

Figure 7.8. Length of a glacier of constant width and a glacier of linearly decreasing width, as calculated with the numerical flowline model. The dashed curve gives the result when a shape factor is included.

Closer inspection makes it clear that in the numerical solution the increase of L with decreasing E is somewhat larger than in the analytical model. This is related to the increasing ice thickness in the numerical model, whereas in the analytical model ice thickness was kept fixed.

It should be noted that the longitudinal glacier profiles for the cases shown in *figure 7.8* differ only slightly. When profiles for glaciers of the same length for a range of values of p are compared, differences are found to be of the order of 5 to 10%. The mean thicknesses differ by only a few percentage points. From this we can conclude that the mean thickness is determined mainly by the glacier length and not so much by the variations in the lateral geometry.

In the final numerical experiment to be discussed in this chapter we look at the effect of the shape factor. When repeating the experiment with linearly decreasing glacier width we expect the lower part of the glacier to be slightly thicker. Here the side drag is more significant and a larger driving stress is needed to obtain ice fluxes that match the mass budget requirements. This implies a larger ice thickness and a positive feedback to the mass balance. So by including the shape factor we expect L to be larger for a given value of E.

The dashed curve in *figure 7.8* shows that this is the case. The effect of the shape factor is larger when the width of the glacier front is smaller. Nevertheless, the effect is restricted and may be not so relevant in view of the other uncertainties in the flowline model (e.g. the choice of the flow parameters, variation in bed characteristics, effect of crevasse fields, distribution of englacial debris, etc.). With regard to real glaciers it will be difficult to demonstrate that the inclusion of a shape factor improves the model simulations, simply because very few data of sufficient detail are available.

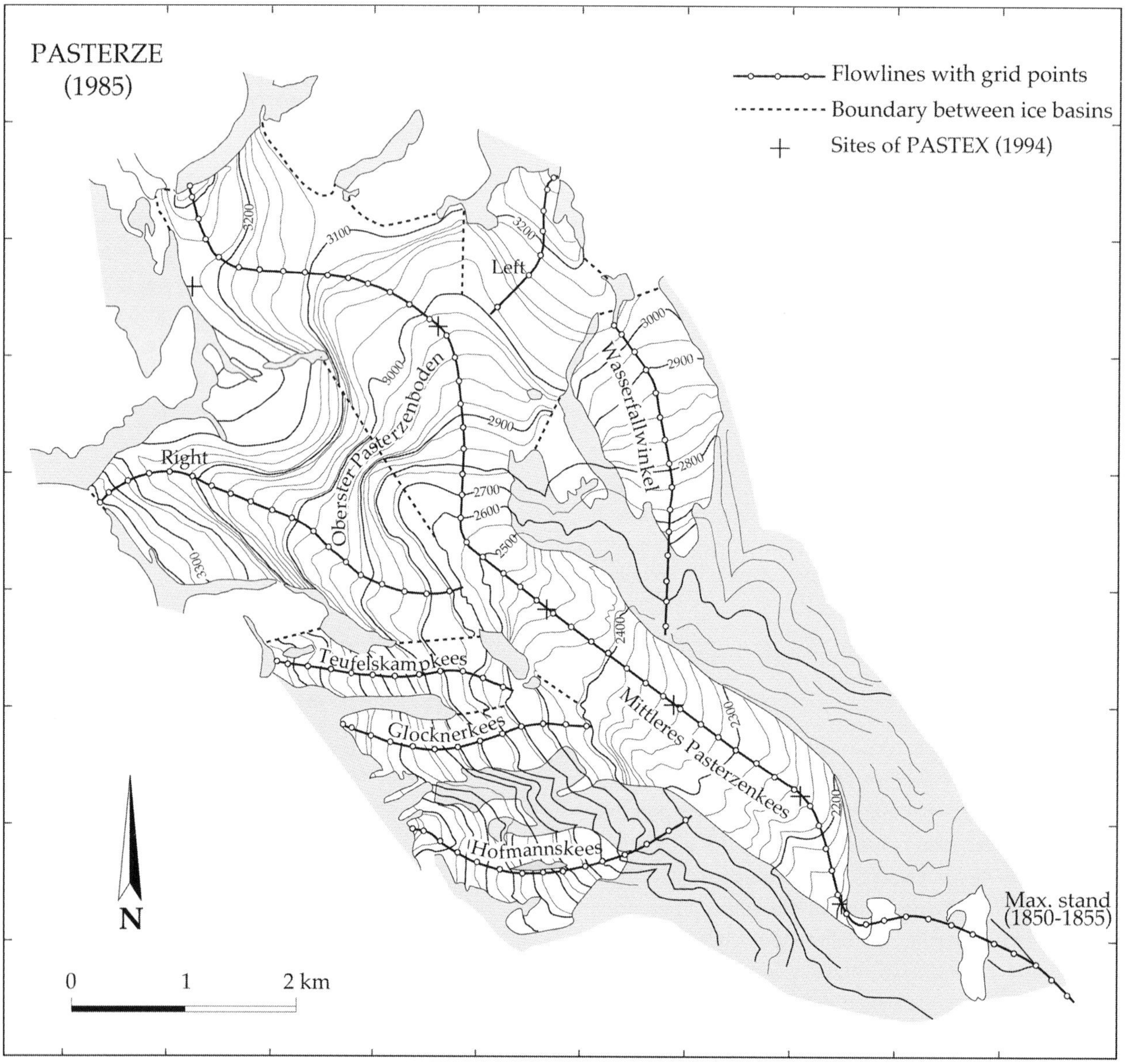

Figure 7.9. The flowline system used to model the Pasterze, Austria [3]. In the 19th century, some of the tributaries (Wasserfallwinkel and Hofmannskees) added mass to the main stream but have now become detached.

7.6 Complex geometry

Many glaciers have a complex geometry. For modelling studies one tries to select glaciers that have a simple geometry, but it is also important that a substantial amount of information is available about changes that have occurred in the past. Unfortunately, a simple geometry and a good historical record do not always come together.

An example is the Pasterze, Austria [3]. An interesting record of historical length variations exist, but the glacier has a complicated geometry. To model this glacier a system of 7 flowlines was used (*figure 7.9*). In the 19th century two of the tributaries, Hoffmannskees and Wasserfallwinkel, added mass to the main stream. Later they became detached to form separate glaciers. Such effects can be handled with the flow-line approach. It is not difficult to construct a scheme for the coupling of a tributary and a main stream. First of all a gridpoint on the main stream is defined that serves as entrance point for the tributary. As soon as the lowest gridpoint of a tributary has a surface elevation equal to that of the entrance gridpoint, ice mass is delivered to the main stream. The amount is determined by the condition that both grid points should have the same surface elevation. The added mass will be diffused in the main stream, but of course its overall dynamics will be affected. Practice has shown that such a coupling scheme behaves more smoothly when the mass added to the main stream is divided over more than one gridpoint (in the case of the model for the Pasterze three grid points were used).

It is also possible to model the dynamics of the tributaries in a more schematic way, as was for instance done in a study of the Unterer Grindelwaldgletscher [4]. The simple analytical models discussed earlier (section 7.2) are quite suitable for this purpose.

References and further reading

1. J.F. Nye (1965): The flow of a glacier in a channel of rectangular, elliptic or parabolic cross-section. *Journal of Glaciology* **5**, 661-690.

2. C.J. van der Veen (1999): *Fundamentals of Glacier Dynamics*. Balkema, Rotterdam, 406 pp.

3 Z. Zuo and J. Oerlemans (1997): Numerical modelling of the historical front variation and the future behaviour of the Pasterze glacier. *Annals of Glaciology* **24**, 235-241.

4. M. J. Schmeits and J. Oerlemans (1997): Simulation of the historical variations in length of the Unterer Grindelwaldgletscher. *Journal of Glaciology* **43** (143), 152-164.

8. Numerical modelling of real glaciers: Nigardsbreen, Rhonegletscher, Abramov Glacier and Franz-Josef Glacier

8.1 From a topographic map to a parameterisation of the 3-D geometry

The analysis of schematic glacier geometries helps us to understand the relative importance of basic parameters like bed slope and balance gradient. But the goal, of course, is to model real glaciers. In this chapter the numerical flowline model will be applied to four glaciers: Nigardsbreen, the Rhonegletscher, Abramov Glacier and Franz-Josef Glacier. Maps of these glaciers are shown in *figure 8.1*. The glaciers are comparable in size, but are quite different with regard to mean slope, area-elevation distribution (see *figure 7.1*) and climatic setting. For Nigardsbreen, the Rhonegletscher and Franz-Josef Glacier a historical length record is available.

First of all flowlines have to be defined. In *figure 8.1* they are indicated by dashed lines; it is clear that some ambiguity is involved. One might just as well use two or three flowlines in the accumulation area, which merge to form a single flowline towards the terminus.

It is very important that the parameterisation of the three-dimensional geometry produces the correct area-elevation distribution. This is not difficult when the bed has been mapped in some detail. However, in most cases the geometry of the bed is not known, or it is known only at a few places. This means that the ice thickness has to be estimated and subtracted from the surface elevation to obtain a bed profile.

The simplest way to estimate the ice thickness from the surface elevation is to assume that the basal shear stress varies little along the flowline and equals the driving stress. As explained in section 5.3 this implies that the product of surface slope and ice thickness is constant:

$$H\left|\frac{dh}{dx}\right| = \frac{S_b}{\rho g} \quad . \qquad (8.1)$$

By choosing a value for the driving stress S_d, one can calculate H from the surface slope. Because of the presence of small-scale features at the glacier surface, which certainly do not obey the simple mechanics adopted for the present glacier model, the surface elevation has to be smoothed. Otherwise eq. (8.1) will produce large variations in ice thickness that have nothing to do with reality.

The driving stress should be chosen with some care. Steeper and larger glaciers have a larger mean basal stress (*figure 6.4*). For most valley glaciers it appears that a value of S_d in the range of $1.5 \cdot 10^5$ and $2 \cdot 10^5$ Pa is a reasonable choice. Once an estimate of ice thickness is available the bed profile follows from

$$b(x_i) = h(x_i) - H(x_i) \quad . \qquad (8.2)$$

Here x_i is the x-coordinate of gridpoint i in the flow direction.

The next step is to determine the parameters λ and w_0 (see *figure 7.5*) for every gridpoint. To obtain an area-elevation distribution in the model that is close to the observed one it is best to determine the glacier width ($w = w_0 + \lambda H$) by following the contour that goes through the gridpoint on the flowline. Combined with an estimate of the steepness of the valley walls this then yields values for λ and w_0.

The fields $b(x_i)$, $\lambda(x_i)$ and $w_0(x_i)$ form the initial geometric data set with which the model can be run. By adjustment of the equilibrium-line altitude the glacier model can be run until a steady state is reached in which the terminus position is at the same location as on the topographic map. However, the simulated surface elevation will not exactly match the surface elevation along the flowline on the map, simply because the model will not produce a steady state with constant basal stress. If the difference is denoted by $\Delta h'(x_i)$, the model can be run again to a steady state with a corrected bed elevation:

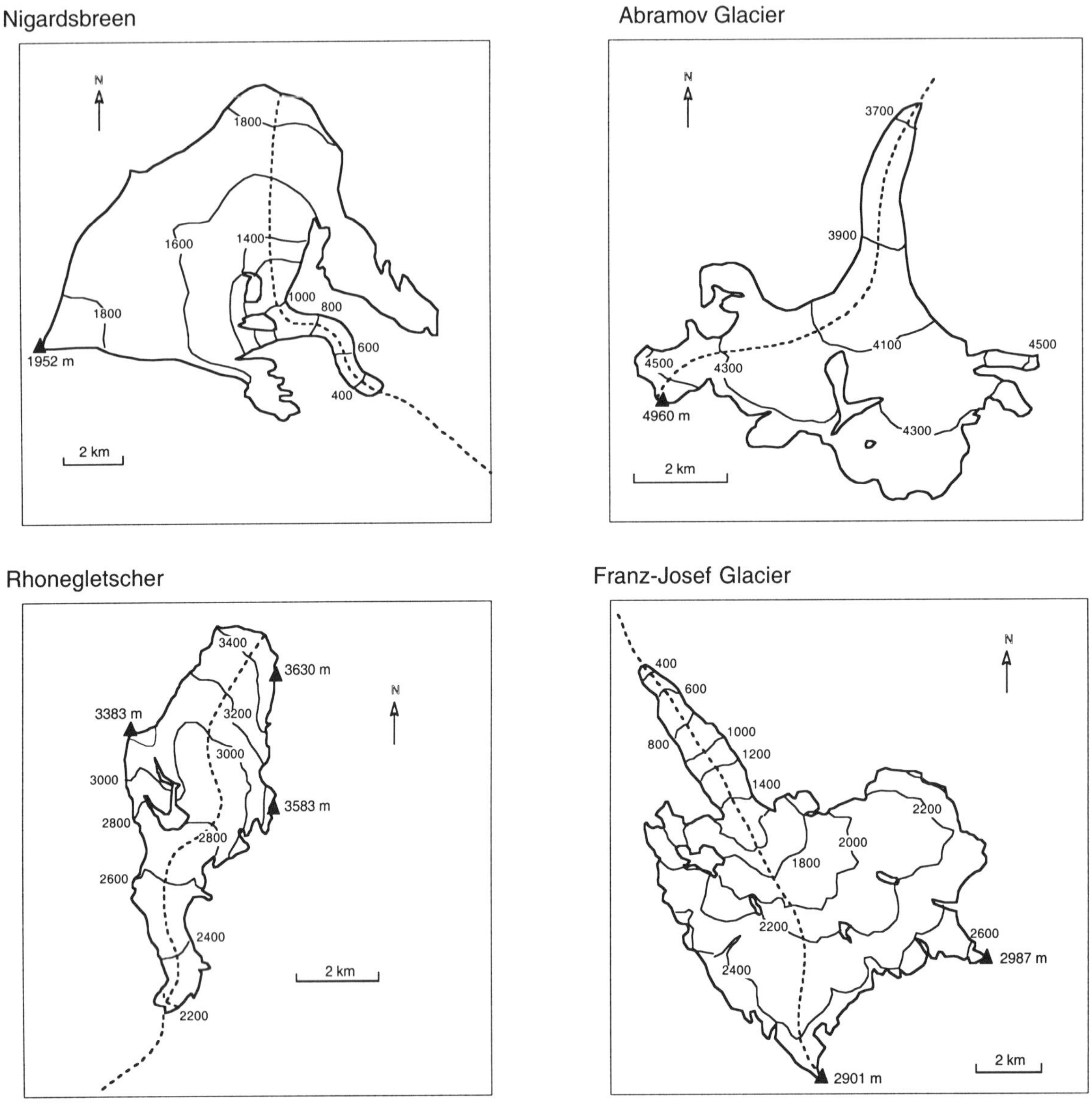

Figure 8.1. Simple maps of the modelled glaciers. Dashed lines are the flowlines as defined for the numerical model.

$$b_{cor}(x_i) = b(x_i) - \Delta h(x_i) \quad . \qquad (8.3)$$

To get the terminus position right E will have to be adjusted. This procedure can be continued until the difference between the simulated and the observed surface elevation at the flowline is sufficiently small.

Next the area-elevation distribution of the model glacier has to be checked. Values of w_0 can be adjusted to obtain the desired accuracy. When the adjustments are large it may be necessary to repeat the tuning procedure for the bed profile.

For glaciers that are relatively smooth the scheme described above works well. We should realise, however, that the assumption of a steady state is generally not valid, and errors will be introduced. For instance, for an inactive and strongly retreating glacier with a flat snout the ice thickness will be overestimated and the bed elevation will be underestimated. Therefore any extra information on the bed geometry is welcome and should be used as a further constraint.

The flow parameters f_s and f_d can also be adjusted. But if no information is available regarding the bed profile, the number of degrees of freedom in the tuning procedure becomes too large. Optimising the flow parameters makes sense if the bed profile is well known. A model glacier that is too thick therefore implies that at least one of the flow parameters is too small. When, over a period of decades, the mass balance and geometric response of a glacier are known it becomes feasible to optimise the flow parameters [1]. But for how many glaciers do we have such information? Anyone carrying out the procedure described above will realise that some ambiguity is involved in the final choice of bed parameters. This does not appear as a serious drawback, however. As argued earlier, the response of glacier length to climatic change on a decadal to century time scale reflects an integral effect in which the details of the geometry do not show up.

8.2 Nigardsbreen, Rhonegletscher, Abramov Glacier, Franz-Josef Glacier

The glaciers shown in *figure 8.1* have been modelled by the author following the procedure described in the previous section. No shape factors are used. Flow parameters are the same for all glaciers, namely those adopted earlier from Budd et al. [2]. The spatial resolution (gridpoint distance) is 100 m. The glacier model is forced by prescribing the balance profile as (z is height):

$$B(z) = B_{ref}(z) + \Delta B + C \quad . \tag{8.4}$$

The reference profile $B_{ref}(z)$ is taken from observations (Nigardsbreen, Rhonegletscher, Abramov Glacier) or from a mass balance calculation (Franz-Josef Glacier [3]), as shown in *figure 8.2*. The perturbation of the balance rate, ΔB, is independent of altitude. In general, the mean observed specific balance is not zero, and the constant C is chosen such that the model produces equilibrium states that are very close to the current glacier stands. This is done to let us make a fair comparison of the climate sensitivity of these glaciers. Later we will discuss the issue of transient behaviour and dynamic calibration.

The bed profiles obtained are shown in *figure 8.3*. The profile of Abramov Glacier is very smooth because it was derived from a schematic map. Indeed, judging from photographs the surface of Abramov Glacier is rather smooth and regular [4]. Clearly, Franz-Josef Glacier is the steepest glacier in this sample and it will be interesting to see what effect this will have on the climate sensitivity.

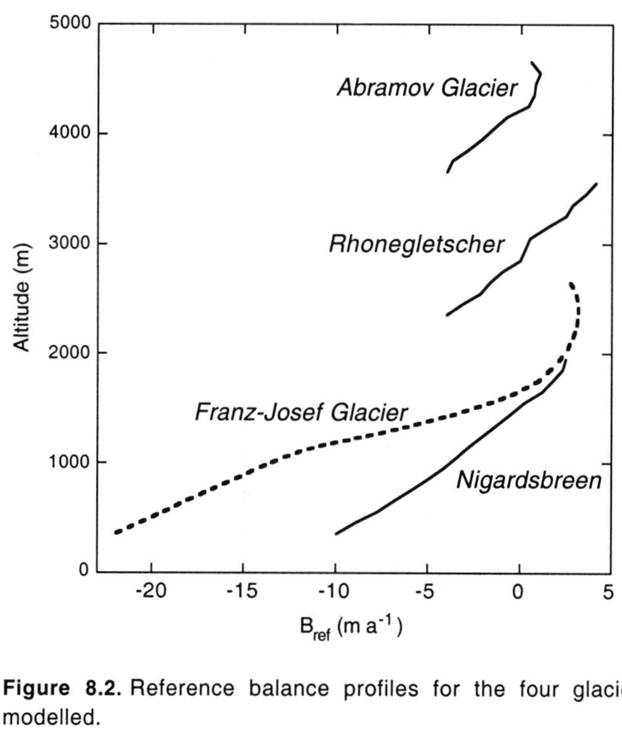

Figure 8.2. Reference balance profiles for the four glaciers modelled.

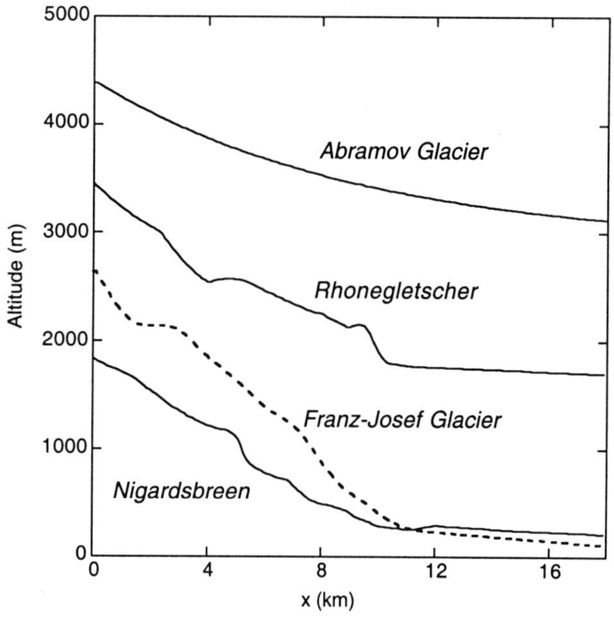

Figure 8.3. Bed profiles for the four glaciers. The profile of Franz-Josef Glacier is dashed for clarity.

A feature common to these glaciers is that the bed slope decreases with x. The little ice-age positions of the glacier fronts were all on the flatter parts of the bed profiles.

In spite of the fact that smoothing has been applied for reasons discussed earlier, two overdeepenings are still present in the bed profile of the Rhonegletscher. In principle such overdeepenings may lead to multiple equilibria [5], but here the magnitude of the overdeepenings is very small and the effect on the overall dynamics is not significant.

Figure 8.4 shows the calculated equilibrium length and volume. The differences between the glaciers are remarkable. The climate sensitivity, interpreted here as the slope of the (L,ΔB)-curve, is large for Nigardsbreen and Abramov Glacier. In contrast, Franz-Josef Glacier is not so sensitive and the Rhonegletscher is in between.

Can we understand the differences in the light of the analytical models developed in Chapters 5 and 6? We used a constant bed slope and a linear balance gradient. So to make at least a qualitative comparison we can estimate these quantites for the glaciers modelled here by making linear fits to b(x) and B(z). The result is shown in *table 8.1*. In a similar way, a characteristic value of the climate sensitivity, defined now as dL/d(ΔB), was derived from the results in *figure 8.4*.

One of the findings of the analysis in section 5 concerned the dependence of the sensitivity on the bed slope s. The change of L with E appeared to be inversely proportional to s. Although in the analytical models we used E and not ΔB as the parameter characterising the climatic conditions, this does not make a difference because the balance gradient was taken to be linear. In this case we have

$$dE = \beta^{-1} d(\Delta B) \, . \tag{8.5}$$

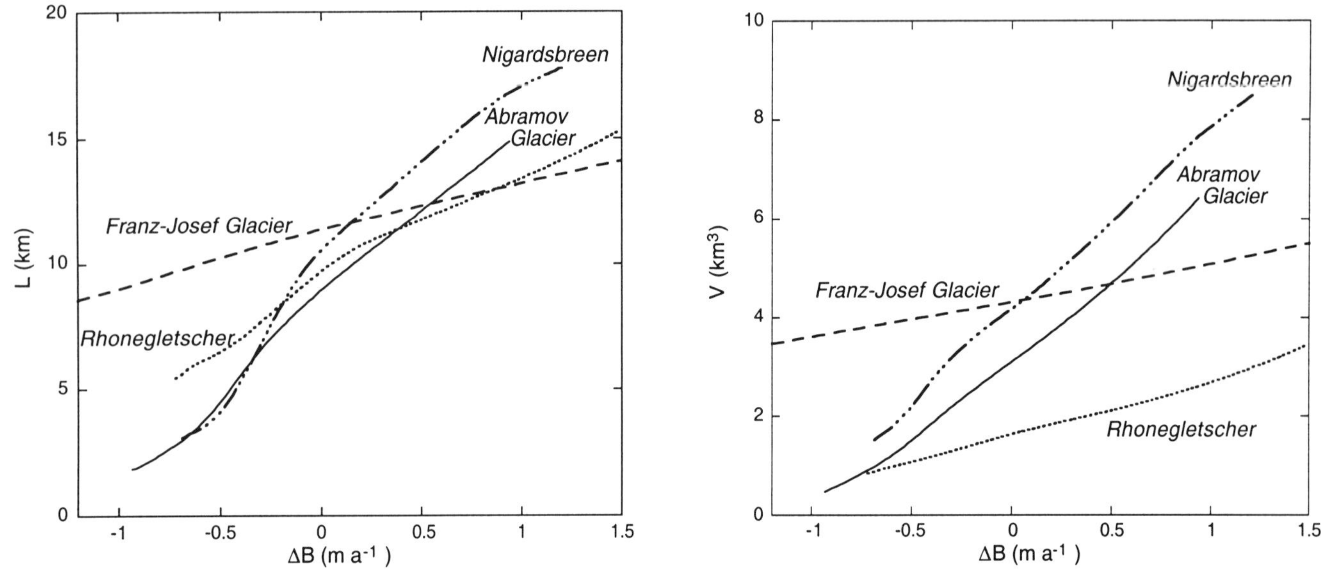

Figure 8.4. Equilibrium length and volume for the four glaciers as a function of the perturbation in the balance rate.

The mean bed slope estimated from a linear fit is denoted by $\tilde{s}$ in *table 8.1*. Clearly, the two most sensitive glaciers have small values of $\tilde{s}$. The Rhonegletscher also has a small slope, but here the effect of lateral geometry is different. We also found (in Chapter 6) that glaciers are more sensitive when the glacier width is decreasing in downflow direction. All four glaciers show this decrease, but it is less pronounced in the case of the Rhonegletscher. In contrast, the width of the tongue of Nigardsbreen relative to the width of the upper basin is very small. This explains why Nigardsbreen is the most sensitive glacier in the sample (and probably one of the most sensitive in the world if we do not consider calving glaciers). It can therefore be concluded that the climate sensitivity values found here from numerical modelling are in qualitative agreement with the simple analytical theory.

Closer inspection of the (L,ΔB)-curves reveals that, except for Franz-Josef Glacier, the sensitivity is by no means constant. There is a range of values for ΔB for which the curves are steeper. It turns out that the largest sensitivity occurs when the narrow part of the valley glacier starts to form, that is, when it is emerging from the wider upper basin. As the smaller part of the glacier continues to grow the sensitivity decreases again. This is in agreement with the theory developed in section 7.2. Again, this effect is most marked for Nigardsbreen, but it is also visible in the curves for Abramov Glacier and the Rhonegletscher.

We should of course realise that for very small values of L the numerical models become less accurate. When a glacier becomes shorter for decreasing values of ΔB, a point will be reached where it breaks up into several smaller glaciers. This is not simulated by the present type of model because it does not contain sufficiently detailed information concerning the glacier bed.

Other interesting bulk quantities are the mean driving stress (S_{dm}) and the mean ice thickness along the flowline (H_m). These are also listed in *table 8.1*. For glaciers of uniform width on a bed of constant slope s the following parameterisation of the mean ice thickness was derived in section 6.4:

$$H_m = \left(\frac{\mu L}{1+\nu s} \right)^{1/2} . \tag{8.6}$$

The values derived for μ and ν were 9 m and 30, respectively. If this equation is applied to the four glaciers, replacing s by $\tilde{s}$, the values found for H_m are slightly too large. A better match is obtained by choosing μ somewhat larger, e.g. a value of 38. The resulting values, denoted by $\tilde{H}$, are also listed in *table 8.1*. Considering the simple form of eq. (8.6) and the complicated geometry of the glaciers studied, the agreement between H_m and $\tilde{H}$ is good.

Driving stresses are in the 1.2 to $2{\cdot}10^5$ Pa range. The value for Franz-Josef Glacier is probably the largest one can expect for valley glaciers. This glacier is fairly steep and has a tremendous mass turnover. The characteristic driving stress can be estimated as (see eq. 5.14)

$$\tilde{S}_d \approx \rho g \tilde{H} \tilde{s} \approx \rho g \tilde{s} \left(\frac{\mu L}{1+\nu \tilde{s}} \right)^{1/2} \tag{8.7}$$

The resulting values for the four glaciers are given in the last column of *table 8.1* (parameter values as above). Again, there is good agreement between the calculated and the estimated values of the driving stress.

At first sight *figure 8.4* suggests that Franz-Josef Glacier is very insensitive to climatic change. However, if the (L,ΔB)-curves are converted to (L,ΔT_{fa})-curves, where ΔT_{fa} is the annual mean change in free atmosphere temperature, the picture changes. Using the results of the mass-balance modelling in section 4.7, *figure 8.5* emerges.

	area (km^2)	L (km)	β (a^{-1})	$\tilde{s}$	$\frac{dL}{d(\Delta B)}$ (km m^{-1} a)	H_m (m)	$\tilde{H}$ (m)	S_{dm}(10^5 Pa)	$\tilde{S_d}$ (10^5 Pa)
Nigardsbreen	48	10.4	0.0078	0.13	8.1	120	125	1.50	1.47
Abramov Glacier	25.9	9.1	0.0055	0.09	7.3	136	135	1.21	1.10
Rhonegletscher	18.5	9.8	0.0066	0.13	4.2	124	121	1.28	1.42
Franz-Josef Glacier	36	11.4	0.0125	0.21	2.1	119	107	2.02	2.02

Table 8.1. Physical characteristics of the modelled glaciers.

Nigardsbreen is still the most sensitive glacier but Franz-Josef Glacier now comes second, which obviously is the consequence of the strong dependence of the mass balance on temperature (*figure 4.15*). Abramov Glacier is the least sensitive glacier, in spite of its small slope. The low sensitivity reflects the more continental character.

A number of other glaciers have been modelled using a similar approach, and it is interesting to compare their sensitivities. *Table 8.2* shows that the differences are significant but certainly not dramatic. In this larger sample too, Nigardsbreen is the most sensitive glacier. A characteristic value of dL/dT_{fa} is 4 km K^{-1}.

So far we have only discussed the bulk properties of the modelled glaciers, but it is also interesting to see how velocity and stress vary along the flowline. We look at two extremes (at least for mid-latitude conditions): Franz-Josef Glacier with its tremendous mass turnover, and Abramov Glacier in a relatively dry continental climate (*figure 8.6*). Not only is there a big difference in mass turnover, but there is also quite a significant difference in bed slope. The variation of the driving stress with x is similar for these glaciers, but the magnitude differs by a factor of two.

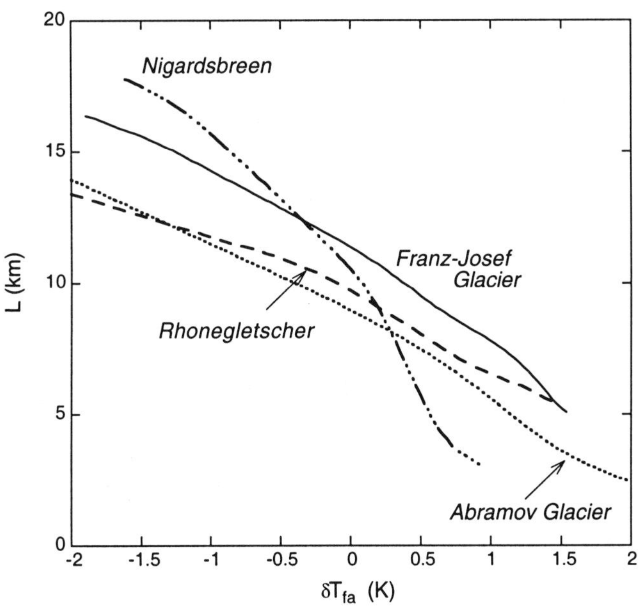

Figure 8.5. Equilibrium glacier length as a function of annual mean air temperature of the free atmosphere.

For a glacier of constant width the highest driving stress is normally found at the equilibrium line. However, when the glacier width decreases we expect the location of maximum driving stress to shift downwards, because a higher velocity is needed to maintain the mass flux.

The ice velocities on Franz-Josef Glacier are impressive for a mid-latitude valley glacier. Unfortunately, there are no data to verify the values calculated, but the values must be fairly accurate if the balance profile (*figure 8.2*) is approximately correct. Sliding contributes a great deal to the ice motion. This is a direct consequence of the parameterisation of sliding. From eq. (5.29) it follows that

$$\frac{U_s}{U_d} = \frac{f_s}{f_d} H^{-2} \ . \qquad (8.8)$$

Sliding, therefore, dominates in the case of a steep glacier like Franz-Josef Glacier. For the set of flow parameters used here, sliding and deformation contribute about equally to the mass flux on Abramov Glacier. When the glacier front is approached, sliding always dominates because the ice thickness goes to zero. Along the flow line, maximum sliding velocities occur in the steep parts, which is certainly realistic for temperate glaciers [6].

glacier	L (km)	$\frac{dL}{dT_{fa}}$ (km K^{-1})
Nigardsbreen	10.4	-6.1
Franz-Josef Glacier	11.4	-3.2
Hintereisferner [7]	7.4	-4.2
Rhonegletscher	9.8	-2.6
Unt. Grindelwaldgletscher [8]	8.5	-4.5
Glacier d' Argentière [9]	10.7	-4.2
Abramov Glacier	9.1	-2.9

Table 8.2. Climate sensitivity of a number of glaciers that have been studied with flowline models. L refers to approximate current glacier length.

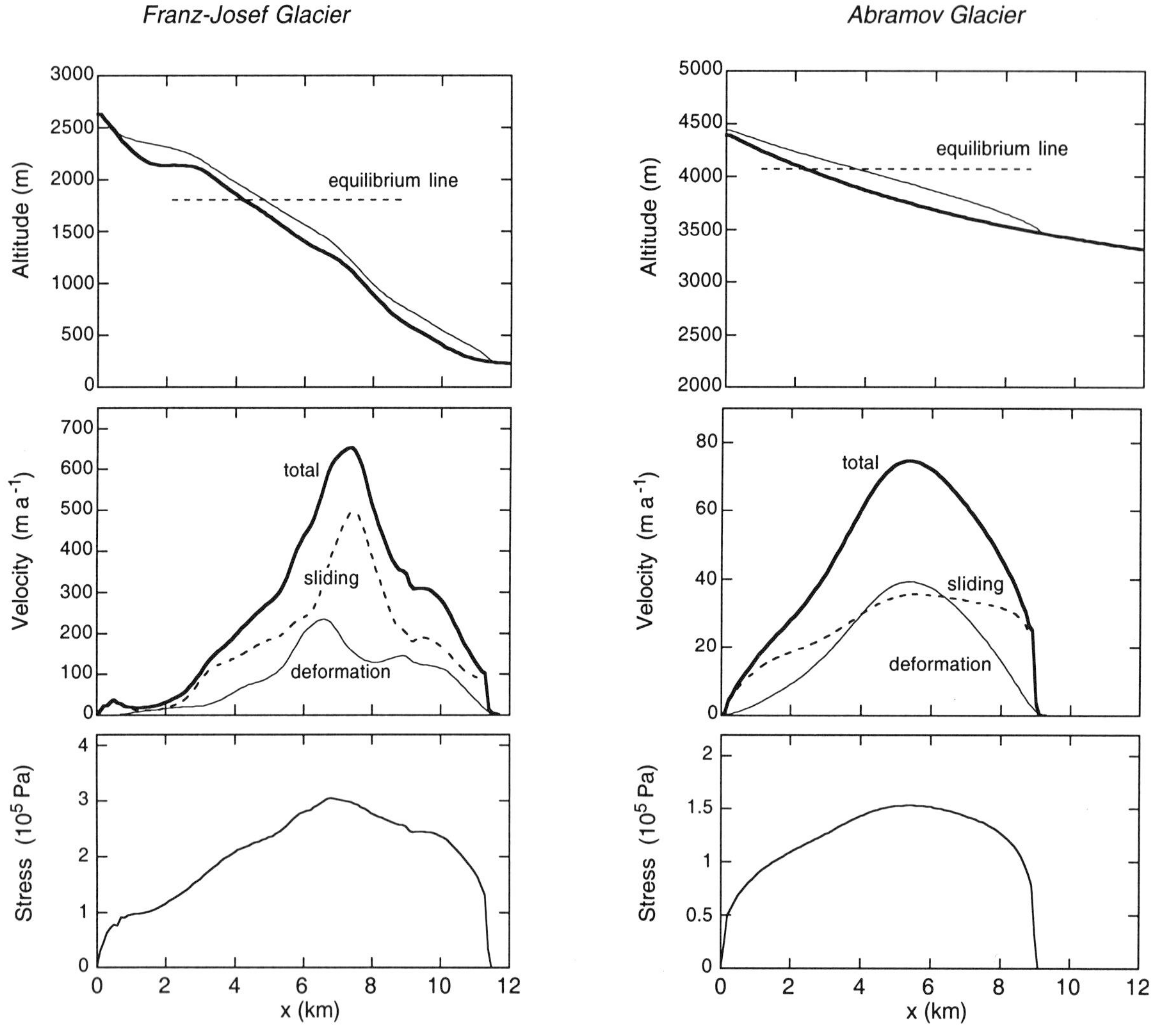

Figure 8.6. Diagnosis of the computed steady states of Franz-Josef Glacier and Abramov Glacier. The figures show the profile, sliding (dashed) and deformational (solid) components of the mean ice velocity in the cross section, and driving stress. The upper two figures are on the same scale to show the differences in profile. The scales of the other figures differ substantially.

8.3 Response times

Response times can be calculated by imposing a stepwise change in the balance rate on a model glacier that is in equilibrium. Because we are dealing with a nonlinear system, the e-folding response time thus found will depend on the magnitude of the change in forcing. The smaller the perturbation of the mass balance, the more linear the response will be. However, too small perturbations lead to changes that cannot be resolved adequately by the grid. So here a balance has to be found.

Figure 8.7 shows the result of a typical numerical experiment to determine the response times for glacier length and volume. Perturbations of +0.4 and -0.4 m a^{-1} in the balance rate have been applied to an equilibrium state of Nigardsbreen. Clearly, the response is not quite symmetric with respect to the sign of the balance perturbation. The glacier adjusts faster to a negative balance perturbation. As discussed earlier in Chapter 6, the volume response is faster than the length response. Response times are also different when other equilibrium states are taken as reference. However, the dependence on factors like balance gradient and bed slope is stronger, so it is still useful to compare the response times of different glaciers.

The e-folding response times for the four glaciers, as defined in section 6.6, are listed in *table 8.3*. Values for positive and negative balance perturbations have been averaged. In spite of its size, Franz-Josef Glacier has a very short response time; this is due to the large balance gradient and the relatively steep bed.

Given the small balance gradient and bed slope of Abramov Glacier, it does not come as a surprise that this glacier responds much more slowly to mass balance perturbations. The Rhonegletscher and Nigardsbreen are in between.

The volume time scale as defined by Jóhanesson et al. [10] was also estimated for the four glaciers. It reads (eq. (6.13)):

$$t_{rV} = \frac{H^*}{-B_{term}} ,$$

	t_V (a)	t_L (a)	$t_{V,Jóh}$ (eq. 6.13)	$t_{L,ana}$ (eq. (6.21)
Franz-Josef Glacier	17	24	6	15
Rhonegletscher	36	58	31	58
Nigardsbreen	42	68	12	48
Abramov Glacier	71	92	35	120

Table 8.3. A comparison of response times for the four glaciers. The two columns on the left are from the output of the numerical model. Values in the other columns are simple estimates according to Jóhannesson et al. [10] and the simple analytical model for a glacier of constant width (Chapter 6).

where H* is a characteristic ice thickness and B_{term} the balance rate at the terminus. The third column in *table 8.3* gives the result (obtained by dividing the mean ice thickness by the ablation rate on the glacier snout). Clearly, the values are much smaller than the volume response time found from the numerical experiments, especially for Franz-Josef Glacier and Nigardsbreen. This is a direct consequence of the large melt rates at the glacier front. Eq. (6.21), which predicts the response time for glacier length, performs better (with an optimal value for c_2), although it tends to exaggerate the differences between the glaciers. The length response time calculated for Abramov Glacier, namely 122 a, appears too large. The response time found for Franz-Josef Glacier, 16 a, is shorter than the result obtained from the numerical experiments. Apparently, then, eq. (6.21) exaggerates the effect of the balance gradient or slope on the response time.

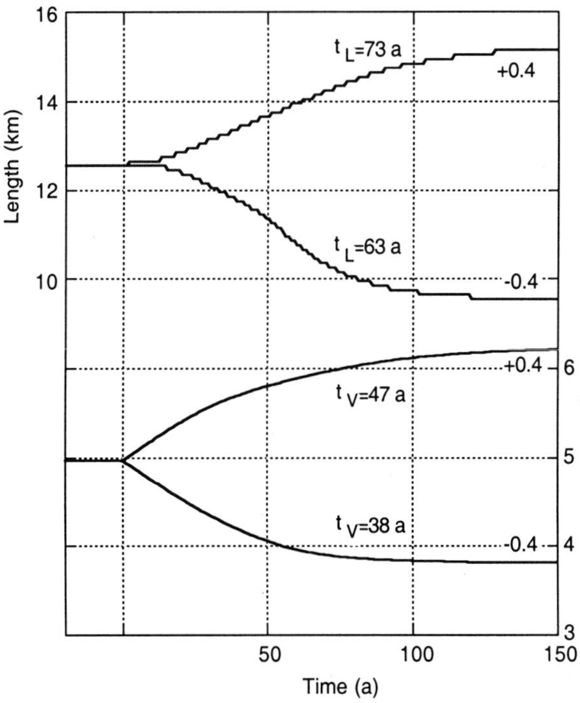

Figure 8.7. Response of Nigardsbreen to a suddenly imposed perturbation of the balance rate of +0.4 and -0.4 m a^{-1}.

8.4 The observed records of glacier length

A considerable amount of information is available regarding the former glacier stands of Nigardsbreen, Rhonegletscher and Franz-Josef Glacier. Length records are shown in *figure 8.8*. Clearly, these glaciers have retreated strongly. In recent years, however, the retreat seems to have come to an end. The climatic interpretation of such records will be discussed in more detail later.

A comparison with *figure 8.5* and *table 8.2* shows that, if the glaciers were not too far from a steady-state, a temperature difference between 0.5 and 1 K would be sufficient to explain the difference in maximum and current length. This is in broad agreement with information from other sources.

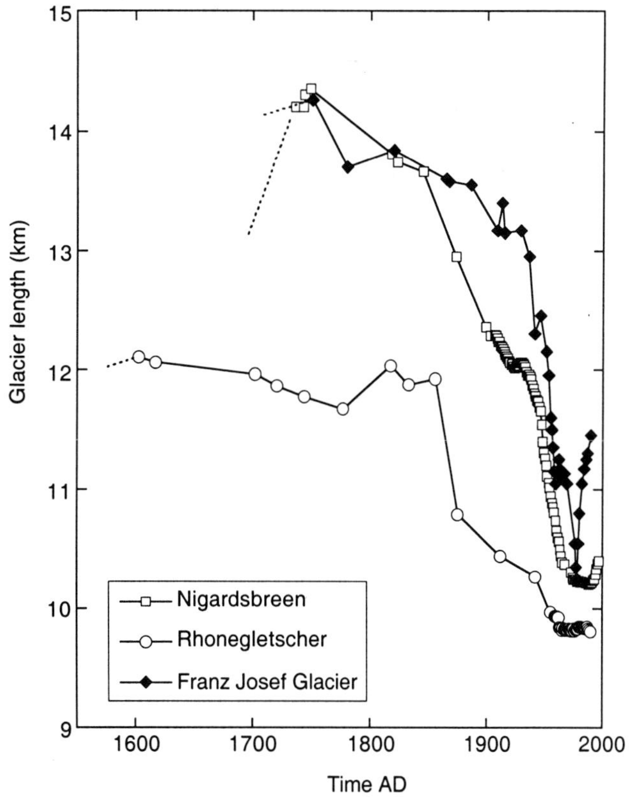

Figure 8.8. Records of glacier length for Nigardsbreen [11], Rhonegletscher [12], and Franz-Josef Glacier [13].

8.5 Dynamic calibration

The fact that glaciers are rarely in a steady state has some implications for the testing of models. Glacier maps always reproduce the observed state, but it is not known to what extent this state is in balance with the current climate. When a reference balance profile has been defined and a proper geometry of the glacier is available, a flow model can be run until a steady state is reached and a comparison of observed and computed ice surface elevation can be made. This is not necessarily a good way to judge the performance of the model, however, because the possible imbalance at the time of comparison is not taken into account.

When a record of glacier length is available the problem can be solved by using this as a constraint for the model simulation. However, this makes sense only if the length of the record exceeds the response time. Forcing the model to reproduce the observed length record is then an effective dynamic calibration. Snapshots from the simulation can be compared with observational evidence (maps, trimlines) to judge the validity of the model.

Dynamic calibration involves the reconstruction of a mass balance history and basically is a way of inverse modelling (reconstruction of a climate proxy from changes in glacier length), which is discussed in more detail later (Chapter 9).

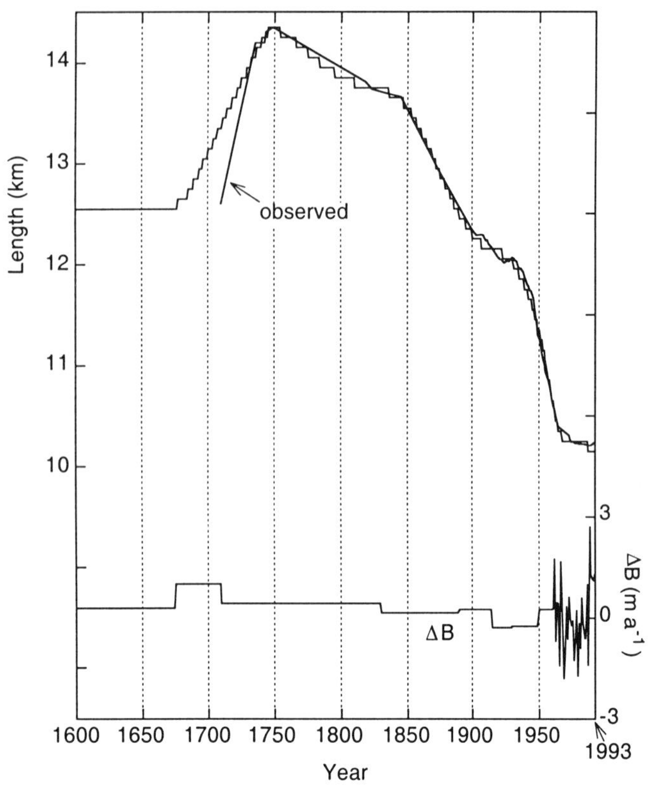

Figure 8.9. Simulation of the historical front variations of Nigardsbreen obtained by adjusting the mass balance perturbation ΔB(t) for the period 1600-1961 (for 1962-1993 the observed values are imposed).

The method was applied by the author to Nigardsbreen [1]. The length record was reproduced by constructing a mass balance history for the period 1600-1961, and using the measured balance values for the period 1962-1993 (*figure 8.9*). ΔB represents a perturbation with respect to the observed balance profile for the period 1962-1993, which was not zero but positive. Consequently, for ΔB = -0.5 m a^{-1}, the mean balance would be zero for the 1993 glacier geometry.

In the historical simulation, the value assigned to ΔB(t) cannot be directly related to the mean specific balance at a certain time, because the balance profiles work out differently for different glacier geometries.

The ΔB(t) curve was constructed in the following way. First ΔB was chosen such that the 1710 glacier stand (L = 12.55 km) represents a steady state. The resulting value of ΔB is +0.28 m a^{-1}. It proved impossible to simulate the very rapid advance during the period 1710-1748 by imposing a realistic perturbation of the balance rate. However, a 10-fold increase in the sliding parameter f_S over a few decades does the job. This results in a very fast advance of the snout, but at the same time the glacier thins so dramatically that a very strong retreat after 1750, much faster than seen in the record, is inevitable. Consequently, this option of simulating the advance by forcing a kind of surge was rejected. When a value of 1 m a^{-1} is taken as the upper limit for ΔB during a period of several decades, the correct 1748 stand can be simulated for ΔB = 1 m a^{-1} for the period 1670-1711, and ΔB = 0.34 m a^{-1} for the period 1711-1760. The simulated rate of advance during the period 1710-1748 is about half of the 'observed' value. A possible conclusion is that the advance rate suggested in a document by Foss (cf. Østrem et al. [11]) is inaccurate. In any case, the timing and stand of the 1748 maximum are indisputable, and the reconstruction of ΔB(t) should be such that this is well simulated by the flow model.

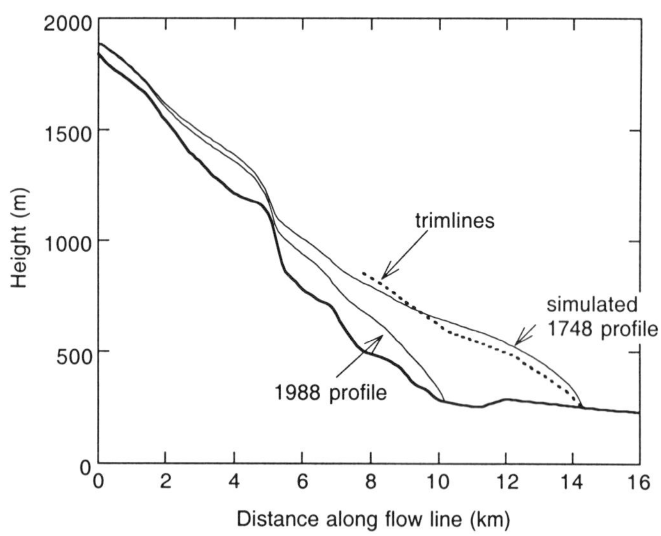

Figure 8.10. A comparison of the simulated and observed surface profiles of Nigardsbreen in 1748.

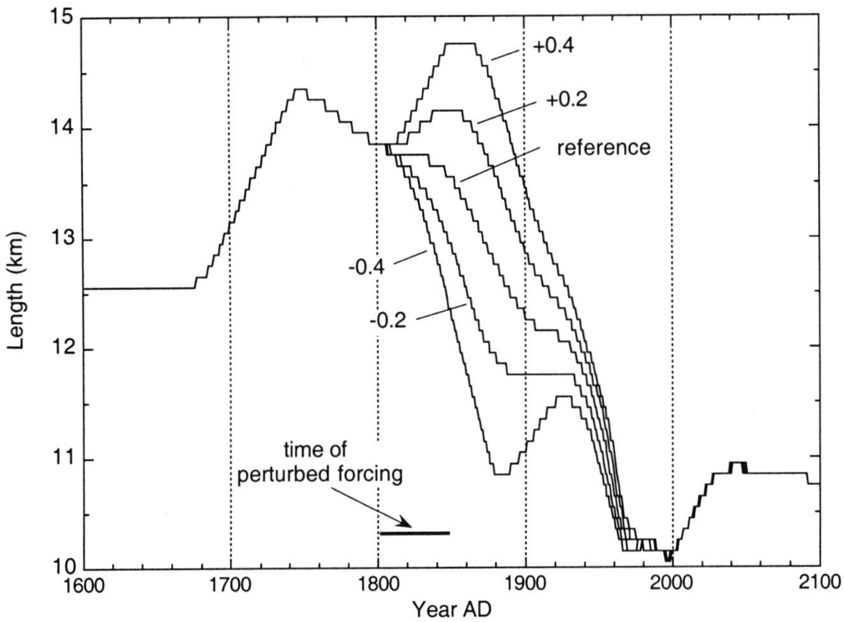

Figure 8.11. Simulated length of Nigardsbreen until 2100, with the mass balance after 1993 equal to the 1962-1993 mean (reference). Also shown are results of experiments in which $\Delta B(t)$ was perturbed in the period 1800-1850 as labelled on the curves (in m a^{-1}).

It is not at all difficult to match the observed length record after 1748. Altogether, 9 step changes in the balance rate were sufficient to obtain a good result. Note that this implies 20 degrees of freedom in the fitting procedure because both the time and the magnitude of the change can be chosen.

In *figure 8.10* the simulated maximum stand is compared to the location of the trimlines. It appears that the simulated profile has a slightly higher elevation than the trimlines. However, there is some uncertainty in a glacier profile derived from trimlines. First of all, a glacier cross profile is not level, but generally shows variations of the order of 10 m. In general, the mean elevation in a profile will be higher than the elevation at the valley walls. On the other hand, the trimlines show maximum stands at particular moments, and it is possible that maximum stands were reached at different times along the profile. In view of these uncertainties, the agreement between the calculated and the inferred surface elevation is good.

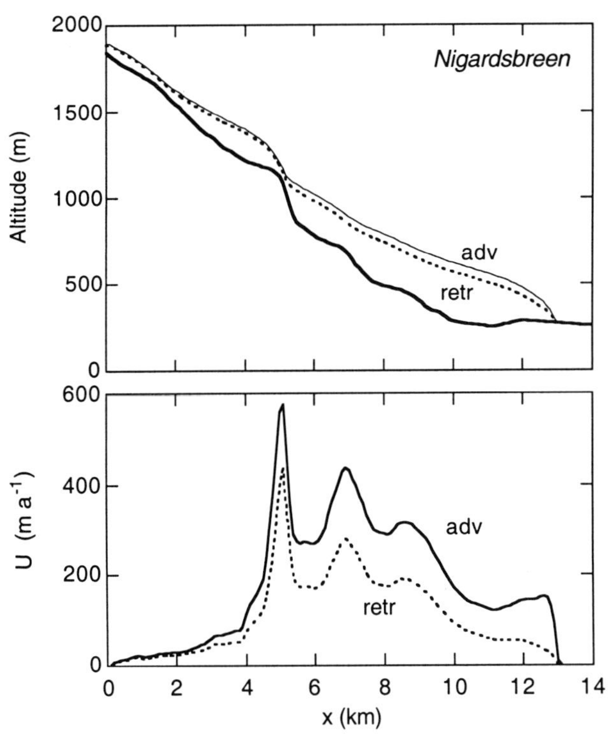

Figure 8.12. A comparison of longitudinal profiles and ice velocity in 1695 (advance, solid) and 1871 (retreat, dashed).

Finally we have a look at how well the mass balance history is determined by the glacier length. *Figure 8.11* shows a set of calculated glacier length curves running until the year 2100. The curve labelled 'reference' is the same as the one shown in *figure 8.9.* Apparently, the model predicts an 800 m increase in length before the glacier is in equilibrium with the 1962-1993 mean balance. The other curves in the figure were obtained by imposing an additional perturbation on the mass balance in the period 1800-1850. The values are given as labels in the figure. Apparently, it takes about 100 years for the effect on the glacier length to disappear. It can also be concluded that the reconstructed mass balance history $\Delta B(t)$ is very well defined: small changes in $\Delta B(t)$ lead to a simulated glacier length record that is unrealistic. So once more we can conclude that glaciers are very good indicators of climatic change!

8.6 Advance and retreat

It is interesting to compare the longitudinal profiles of Nigardsbreen during advance and retreat, because such a comparison provides an idea of how a relation between mean ice thickness and glacier length is violated during periods of rapid change. *Figure 8.11* shows surface elevation at the time when Nigardsbreen was 13 km long: in 1695 during strong advance and in 1871 during fast retreat. There is a clear

difference between the profiles. The mean ice thickness along the flowline is 196 m during the advance and 164 m during the retreat, i.e. a difference of 32 m. With regard to the overall mean ice thickness the difference is smaller (26 m). A mass balance gradient of 0.0077 a^{-1} increases the effective mass balance perturbation by 0.2 m a^{-1}, which is a significant amount (compare *figure 8.4*). So, as far as Nigardsbreen is concerned, the feedback between changes in surface elevation and mass balance cannot be neglected.
The volumes during advance and retreat are 6.0 and 5.1 km^3. This is mainly a consequence of the differences in ice thickness, but the slightly larger area of the glacier in advancing mode (wider tongue) also makes a contribution.
The second panel in *figure 8.12* shows the ice velocity for the two states. The differences are small in the upper basin, but larger further down. Close to the ice front, around x = 12 km, the velocity in the advancing state is much higher than in the retreating state, as expected.

References and further reading

1. J. Oerlemans (1997): A flow-line model for Nigardsbreen: projection of future glacier length based on dynamic calibration with the historical record. *Annals of Glaciology* **24**, 382-389.

2. W.F. Budd, P.L. Keage and N.A. Blundy (1979): Empirical studies of ice sliding. *Journal of Glaciology* **23**, 157-170.

3. J. Oerlemans (1997): Climate Sensitivity of Franz-Josef Glacier, New Zealand, as revealed by numerical modelling. *Arctic and Alpine Research* **29** (2), 233-239.

4. Glacier Mass Balance Bulletin No. 4. World Glacier Monitoring Service. IAHS (ICSI) - UNEP - UNESCO, 1996.

5. J. Oerlemans (1989): On the response of valley glaciers to climatic change. In: *Glacier Fluctuations and Climatic Change.* (ed.: J. Oerlemans), Reidel, 353-371.

6. R. LeB. Hooke (1998): *Principles of Glacier Mechanics.* Prentice Hall, New Jersey, 248 pp.

7. J.W. Greuell (1992): The Hintereisferner: mass-balance reconstruction and numerical modelling of the historical length variations. *Journal of Glaciology* **38** (129), 233-244.

8. M. J. Schmeits and J. Oerlemans (1997): Simulation of the historical variations in length of the Unterer Grindelwaldgletscher. *Journal of Glaciology* **43** (143), 152-164.

9. Ph. Huybrechts, P. de Nooze and H. Decleir (1989): Numerical modelling of Glacier d'Argentière and its historic front variations. In: *Glacier Fluctuations and Climatic Change.* (ed.: J. Oerlemans), Reidel, 373-389.

10. T. Jóhannesson, C.F. Raymond and E.D. Waddington (1989): Time-scale for adjustment of glaciers to changes in mass balance. *Journal of Glaciology* **35**, 355-369.

11. G. Østrem, O. Liestøl and B. Wold (1977): Glaciological investigations at Nigardsbreen, Norway. *Norsk Geogr. Tidsskr.* **30**, 187-209.

    E. Elvehøy and N. Haakensen (eds) (1992): *Glasiologiske Undersøgelser i Norge 1990 og 1991.* Norges Vassdrags- og Energiverk, Oslo, 103 pp.

    N. Haakensen (ed) (1995): *Glasiologiske Undersøgelser i Norge 1992 og 1993.* Norges Vassdrags- og Energiverk, Oslo, 139 pp.

12. J. Wallinga and R.S.W. van de Wal (1998): Sensitivity of the Rhone glacier to climate change: experiments with a one-dimensional flow-line model. *Journal of Glaciology* **44** (147), 383-393.

13. V. Brazier, I.F. Owens, J.M. Soons and A.P. Sturman (1992): Report on the Franz Josef Glacier. *Zeitschrift für Geomorphologie* **86**, 35-49.

    T.J.H. Chinn (1989): Glacier of New Zealand. In: *Satellite Image Atlas of Glaciers in the World; Irian Jaya, Indonesia, and New Zealand.* U.S. Geological Survey Professional Paper 1386-H.

    P. Kasser (1973) *Fluctuations of Glaciers, 1965-1970.* (Vol. II) IASH (ICSI)- UNESCO.

    F. Müller (1977) *Fluctuations of Glaciers,1970-1975.* (Vol. III) IASH (ICSI)- UNESCO.

    W. Haeberli (1985): *Fluctuations of Glaciers, 1975-1980.* (Vol. IV) IASH (ICSI)- UNESCO.

    W. Haeberli and P. Müller (1988): *Fluctuations of Glaciers, 1980-1985.* (Vol. V) IASH (ICSI)- UNEP-UNESCO.

    W. Haeberli and M. Hoelzle (1993): *Fluctuations of Glaciers, 1985-1990.* (Vol. VI) IASH (ICSI)- UNEP-UNESCO.

    W. Haeberli, M. Hoelzle, S. Suter and R. Frauenfelder (1998): *Fluctuations of Glaciers, 1990-1995.* (Vol. VII) IASH (ICSI)-UNEP- UNESCO.

9. Climatic interpretation of glacier fluctuations

9.1 Introduction

In the preceding chapters concepts have been introduced and models developed that should enable us to make a climatic interpretation of glacier records. There is a wealth of data on glacier fluctuations, in particular glacier length. A selection of records is shown in *figure 9.1*. The nature of the data differs considerably from glacier to glacier and from period to period. In recent decades surveying methods have been applied to many glaciers; earlier information comes mainly from terminal moraines, old photographs, paintings and etchings. *Figure 9.2* shows an illustration of the Rhonegletscher, Switzerland.

These records make it clear that glacier retreat is a world-wide phenomenon. Nevertheless, there are large differences in the distance over which glacier fronts have retreated. Sometimes the long-term changes are obscured by rapid changes that are not due to climatic forcing. The record of Leirufjardarjökull, Iceland, for instance, shows steady retreat interrupted by a surge [1]. The statement that surging glaciers are not very useful as climate indicators is basically true, but this depends on the time scale. If records are sufficiently long surging glaciers can nevertheless provide useful information. Several Icelandic records show long-term trends on which surges are imposed (implying also that the surging behaviour of at least some glaciers is retained when they get smaller).

The records in *figure 9.1* do not cover the globe well. It is probably true to say that 95% of all data on glacier length comes from Europe. However, there is a tremendous overlap in this information because neighbouring glaciers show very similar behaviour on longer time scales. The problem of poor spatial coverage over the globe has of course been recognised for a long time, and efforts are now being made to improve the situation [2]. Clearly, satellite data can play an important role here.

Climatic interpretation of glacier length records can be made in many ways. In regions where the front fluctuations of many glaciers have been observed on a regular basis, the information can be treated in a statistical way. In some studies the fraction of retreating or advancing glaciers has been interpreted as a climate proxy. Clearly, however, this leads to a loss of information because it is implicitly assumed that all glaciers have the same response time. The use of individual glacier records for climate reconstruction was first given a mathematical basis by Nye [3], but the method has hardly been used.

As a kind of intermediate approach, the author has performed a statistical analysis of retreat rates using a minimum of physics [4]. In this study changes in glacier length were corrected for differences in sensitivity related to the mean slope and the climatic setting. However, response times were not considered, and only the long-term linear trend was studied.

Another possibility is to use flowline models. The procedure of dynamic calibration discussed in section 8.5 produces a mass-balance history, which is a climate proxy. Flowline models seem to be the most appropriate models, because these take geometric effects on the sensitivity and response time of a particular glacier into account. However, a great deal of information on glacier length comes from glaciers that have not been mapped in any detail. Nevertheless, such information is very valuable because it usually comes from regions with short meteorological records. There is therefore room for the development of simpler models that take into account the bulk characteristics of a glacier. An interesting study of this kind was performed by Haeberli and Hoelzle [5].

All efforts to extract climatic information from glacier records can be classified as inverse modelling. The glacier is a physical system, and given the response it is up to us to determine the forcing. So it is useful to begin with a general discussion about the application of inverse modelling.

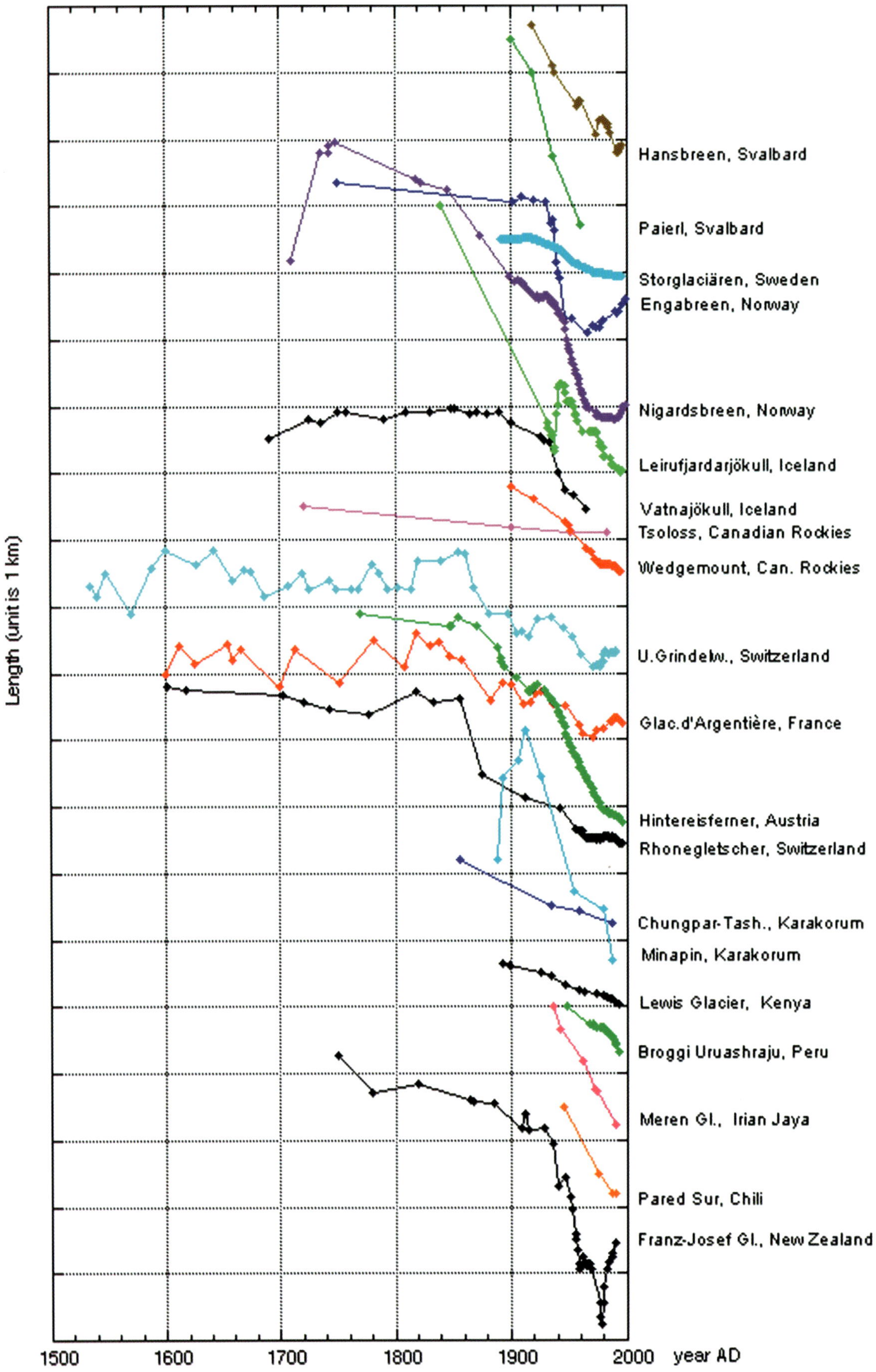

Figure 9.1. Glacier length records from different parts of the world. Symbols refer to individual data points, which have been connected by straight lines (not necessarily implying that there was little fluctuation between two data points). Data from the World Glacier Monitoring Service with some additions.

Figure 9.2. Retreat of the Rhonegletscher, Switzerland. The upper picture shows the state of the glacier around 1800, shortly before it reached its maximum stand. The middle picture shows the situation in 1894. The modern photograph was made in the 1980s.

9.2 Inverse modelling

We consider a linear system and define a reference state (L_0, E_0) by writing

$$L_0 = L(t) - L'(t)\ ;\ E_0 = E(t) - E'(t)\ . \qquad (9.1)$$

The change in glacier length is then related to fluctuations in the equilibrium-line altitude by

$$\frac{dL'(t)}{dt} = -\frac{\{cE'(t) + L'(t)\}}{t_{rL}}\ . \qquad (9.2)$$

Here t_{rL} is the length response time, t time, and the constant c determines how the steady-state glacier length is related to the equilibrium-line altitude.

One well-known way of solving this equation is to use a Fourier-transform in which L' and E' are effectively resolved in periodic functions in time. This procedure is described in nearly every textbook on differential equations.

Here we follow a different route. First of all we solve eq. (9.2) for E':

$$E'(t) = -\frac{1}{c}\left[L'(t) + t_{rL}\frac{dL'(t)}{dt}\right]\ . \qquad (9.3)$$

It is not possible to apply this equation directly to glacier length records. A record has to be prepared for further analysis. We do this by linear interpolation between observed values and then fit a polynomial of degree N to the resulting data with a least-squares method. Therefore

$$L'(t) = \sum_{n=0}^{N} p_n t^n\ . \qquad (9.4)$$

Here the p_n are the coefficients. The time derivative is then simply

$$\frac{dL'(t)}{dt} = \sum_{n=1}^{N} n\, p_n\, t^{(n-1)}\ . \qquad (9.5)$$

So once the p_n have been determined and the response time and sensitivity of the glacier are known it is easy to reconstruct the equilibrium-line altitude using eq. (9.3).

The choice of N determines the level of detail obtained in the reconstructed E'(t) curve. It is not advisable to take a high value, however. First of all it is not realistic to assume that frontal fluctuations with a small time scale always reflect changes in climate on a regional or even larger scale. Secondly, in most glacier records data points are sparse in the earlier periods, but from a climate reconstruction point of view these are the most interesting periods.

As an example we consider the record of the Rhonegletscher. The data points are rather sparse, except for the last 50 years (*figure 9.3*). In this case a polynomial with $N = 4$ was used, which gives an acceptable fit. Values used for the sensitivity c and response time t_{rL} are 32 and 58 a, respectively. These values were derived from experiments with the flowline model for the Rhonegletscher (Chapter 8). The calculated E'(t) curve is shown in *figure 9.3*, together with the polynomial fit to the observed length record. For comparison, curves for a doubled and halved response time are also shown.

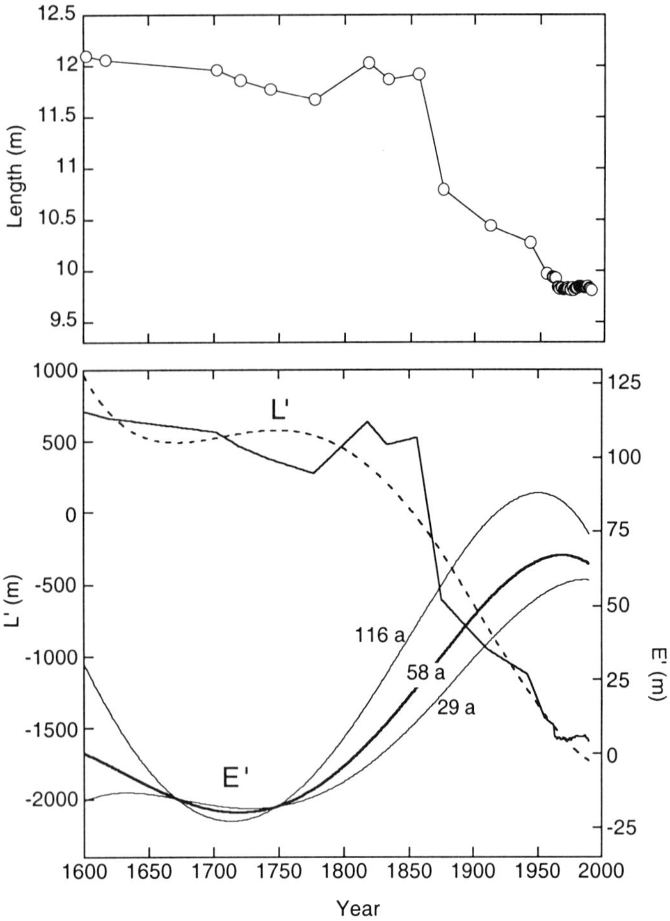

Figure 9.3. Reconstruction of E' from the length record of the Rhonegletscher (upper panel). The L'-curve is a fourth order polynomial fit. Labels on the E'-curves give the value of the response time used.

It turns out that the rise of the equilibrium line needed after the little ice-age to explain the retreat of the Rhonegletscher is of the order of 100 m. A longer response time shifts the E'-curve backwards in time and makes the amplitude larger.

It is clear that the use of polynomials simplifies the analysis. However, a (minor) disadvantage is that the mean value of the calculated E' is not necessarily zero over the time span considered. It is more elegant to use orthogonal functions, like Legendre-polynomials or Fourier-functions, to represent the observed length record. Such functions provide more scope for manipulation with regard to the separation of time scales. This remains a topic for further research.

	c	t_{rL} (a)
Unterer Grindelwaldgletscher [6]	40	40
Rhonegletscher [7]	32	58
Glacier d'Argentière [8]	35	32

Table 9.1. Climate sensitivity and response time used in reconstructing E' for three glaciers in the Alps.

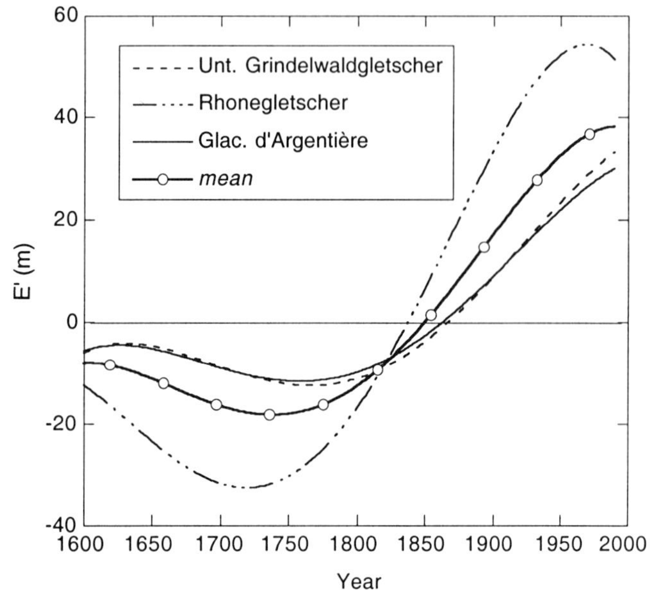

Figure 9.4. Reconstructed changes in equilibrium-line altitude for three glaciers in the Alps.

As a further example of inverse modelling with polynomials *figure 9.4* shows the result obtained for the Unterer Grindelwaldgletscher and the Glacier d'Argentière. In all cases N = 4 and the values for the sensitivity and response time were taken from *table 9.1*. A (small) constant was added to the calculated E'-curves to make the mean zero for the period 1600-1990. All the E'-curves show a rise of the equilibrium-line altitude, of course, but the rise seems to be larger for the Rhonegletscher. The mean of the three E'-curves is also shown and provides an estimate of the change in equilibrium-line altitude in the Alps on a 100-year time scale.

The linear model is easy to use, but may be not sufficiently accurate when a large range of glacier lengths is considered. A second-order formulation for the equilibrium glacier length of the type

$$L'_{eq} = c_1 E' + c_2 E'^2 \qquad (9.6)$$

may then be preferable. Again, c_1 en c_2 are constants that have to be determined from a glacier model. Using eq. (9.5) then leads to a quadratic equation that can be solved easily. It reads

$$c_2\, E'(t)^2 + c_1\, E'(t) + \sum_{n=0}^{N} p_n\, t^n + t_{rL} \sum_{n=1}^{N} n\, p_n\, t^{(n-1)} = 0 \qquad (9.7)$$

The basic assumption in the method described in this section is that the response of the glacier length is linearly damped in time. This is an approximation that can be valid only for changes that are neither too large (otherwise the response time would change significantly because the shape of the glacier changes) nor too small (because then the details of the local mechanics come into play). With regard to glacier

fluctuations over the past centuries, it appears that these conditions are fulfilled in many cases. Numerical experiments with flowline models have shown that the response times for positive and negative mass balance perturbations are normally not the same, but the differences are not so large that the simple inverse method is inapplicable. The method can be applied easily to glaciers for which few data are available. Combining it with one of the simple analytical glacier models discussed earlier, or with the analysis given in the next section, provides a tool with which a large number of glaciers can be handled.
Eq. (9.6) can be further refined. In fact, any functional relation, or look-up table, between L'_{eq} and E' can be handled. The response time can even be allowed to vary with L.

9.3 Another look at glacier fluctuations of small amplitude

In the approach described in section 9.2 the response of glacier length to climate change was assumed to be a linear process in which the parameters (c and t_{rL}) have to be determined by a (simple or complex) glacier flow model. It is also possible to arrive at a linear model by performing a perturbation analysis on the continuity equation. This yields explicit formulations for c and t_{rL}.
Conservation of ice volume V can be expressed as

$$\frac{dV}{dt} = \frac{d(A\bar{H})}{dt} = \bar{H}\frac{dA}{dt} + A\frac{d\bar{H}}{dt} \quad . \qquad (9.7)$$

Here A is the glacier area and $\bar{H}$ the mean ice thickness. Now we write $V = V_0 + V'$, $A = A_0 + A'$ and $\bar{H} = \bar{H}_0 + \bar{H}'$, where the reference state is defined as $(V_0, A_0, \bar{H}_0)$. By substitution in eq. (9.7) and by neglecting higher-order terms we then obtain the perturbation equation

$$\frac{dV'}{dt} = \bar{H}_0\frac{dA'}{dt} + A_0\frac{d\bar{H}'}{dt} \quad . \qquad (9.8)$$

The next step is to relate the change in glacier area to the change in glacier length. We assume

$$A' = w_f L' , \qquad (9.9)$$

where w_f is the typical width of the glacier snout. We also assume that the change in mean thickness is proportional to the change in glacier length (linearised form of eq. (8.6)):

$$\bar{H}' = \eta L' \quad . \qquad (9.10)$$

So for the change of ice volume with time we find

$$\frac{dV'}{dt} = (\eta A_0 + w_f \bar{H}_0)\frac{dL'}{dt} = \text{change of ice volume} \quad . \qquad (9.11)$$

Now we have to find an expression for the actual ice volume added or removed from the glacier. In a perturbation analysis we have three contributions:

$$\text{change of ice volume} = A_0 B' + A_0 \beta \bar{H}' + w_f B_f L' \quad . \qquad (9.12)$$

B' is the perturbation of the balance rate (constant over the glacier), β the balance gradient and B_f the characteristic balance rate at the glacier front. The first term is simply the amount of ice added or removed by the balance perturbation on the reference glacier area. The second term represents the feedback between balance rate and surface elevation, and, by using eq. (9.10), can be expressed in L'. The third term is the amount of ice lost or gained because the length of the glacier may deviate from that of the reference state. Combining eqs. (9.11)-(9.12) yields

$$\frac{dL'}{dt} = \frac{A_0}{\eta A_0 + w_f \bar{H}_0} B' + \frac{\eta \beta A_0 + w_f B_f}{\eta A_0 + w_f \bar{H}_0} L' \quad . \qquad (9.13)$$

This equation is of the same form as eq. (9.2), but now we have derived the coefficients explicitly. The equilibrium state is obtained by setting the time derivative to zero:

$$L' = -\frac{A_0 B'}{\eta \beta A_0 + w_f B_f} \quad . \qquad (9.14)$$

The response time for glacier length according to this model is apparently

$$t_{rL} = -\frac{\eta A_0 + w_f \bar{H}_0}{\eta \beta A_0 + w_f B_f} \quad . \qquad (9.15)$$

In the absence of feedback between glacier surface elevation and mass balance ($\eta = 0$), the response time reduces to $-\bar{H}_0/B_f$, which corresponds to the volume time scale as defined by Jóhannesson ([9], eq. (6.13)). In this case eq. (9.14) reduces to

$$L' = -\frac{A_0 B'}{w_f B_f} \quad , \qquad (9.16)$$

which is similar to the equation used by Haeberli and Hoelzle [5]. So eq. (9.13) is a slightly more general linearisation because it takes into account the altitude - mass balance feedback.
Figure 9.6 shows that this feedback is really significant. The response time is plotted as a function of η. From *figure 6.6* it follows that a typical value of η is 0.005. Depending on the glacier area and width of the glacier front, this may easily double or triple the response time when compared to the case with $\eta = 0$.
In Chapter 6 it was found that the response time is larger for glaciers on a bed with a small slope s. Implicitly, this is also

the case for t_{rL} as defined by eq. (9.15). As shown in *figure 6.6* and formulated in eq. (6.3), for smaller slopes the value of η is larger, resulting in larger values of the response time.

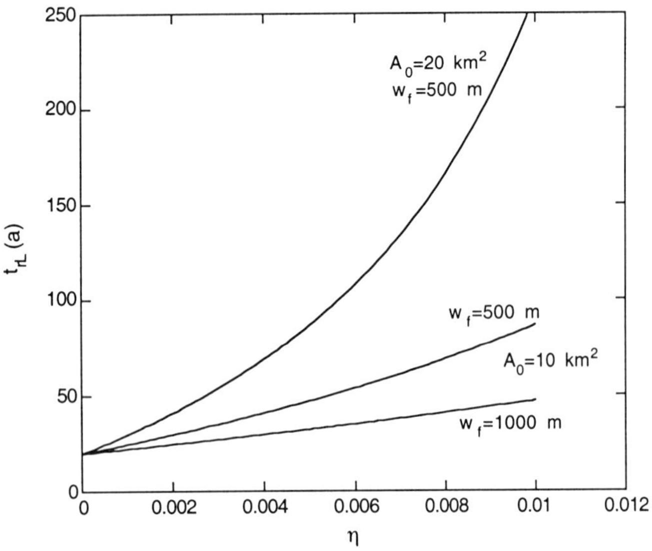

Figure 9.5. Response time for glacier length calculated from eq. (9.15). The balance rate at the glacier front was set to -5 m a^{-1}, the balance gradient to 0.007 a^{-1}. Values for the glacier area and width of the glacier front are given as labels on the curves.

9.4 Inverse modelling with flow-line models

Because of the inherent nonlinearity of numerical flowline models, their use in inverse modelling is less straightforward. We cannot simply reverse the time axis because hysteresis effects are easily generated within the model dynamics. As a consequence, a variational method has to be used to reconstruct a mass-balance history from a glacier length record. In such an approach values of control parameters have to be determined that minimise the discrepancy between data (observed glacier length, L_{obs}) and model results (calculated glacier length, L_{cal}).
We define the quantity to be minimised as

$$D_L = \left\{ \frac{1}{K} \sum \left(L_{cal} - L_{obs}\right)^2 \right\}^{1/2} . \qquad (9.17)$$

The summation is over all K annual values. Because data are not generally available for every year, interpolation has to be used to fill the gaps (unless we do not want to give equal weights to equal periods). The set of control parameters will normally consist of a climate proxy (for instance a mass balance perturbation), but may also include other quantities like flow parameters or geometric characteristics. However, the number of control parameters should be compatible with the degree of detail in the data used to minimise D_L.

The simplest approach is one in which only a mass-balance perturbation ΔB, independent of altitude, is used as a control parameter. In fact ΔB is a control vector, which may consist of a succession of step functions where the number of values for ΔB_i is prescribed. Denoting this number by N, we have 2N unknowns (including the times t_i at which the balance perturbation takes on the value ΔB_i). A straightforward way is to start with a good initial guess, where N is determined by the character of the curve of the observed glacier length and the degree of detail required (or, rather, that appears meaningful in the light of other model uncertainties). A steepest descent method can be applied to minimise D_L with respect to the parameter vector $\mathbf{P}=[t_1, \Delta B_1, ..., t_N, \Delta B_N]$. In that case a (very) large number of runs have to be carried out to determine the maximum of grad(**P**) in the parameter space.
In practice it is more feasible to vary ΔB_i and t_i for each i separately and progressively in time, starting from a good initial state of **P** (this is not difficult when a characteristic response time and steady-state solutions have been calculated first as a function of ΔB). So D_L can be minimised by determining the optimum values of the pairs $[\Delta B_i, t_i]$ independently.
The reconstruction of the mass balance history of Nigardsbreen, presented as 'dynamic calibration' in section 8.5, has been carried out in this way. A good result was obtained for N=9, i.e. 9 pairs of values for $[\Delta B_i, t_i]$ were sufficient to obtain a fairly accurate match of the simulated and the observed front positions for the period 1748 - 1993 (*figure 8.9*).

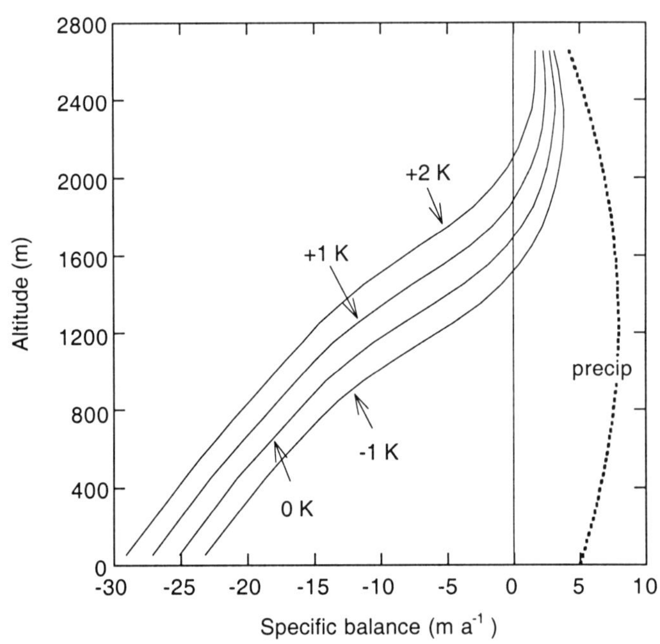

Figure 9.6. Balance profiles for Franz-Josef Glacier as calculated with a mass balance model [10]. Labels on the curves give a uniform temperature change with respect to a reference case. The precipitation curve (input) is also shown.

So far we have taken as a control parameter the equilibrium-line altitude (E') or a uniform mass-balance perturbation (ΔB). However, we can perform a more refined analysis by dealing with the fact that balance profiles may change as well. As an example we consider Franz-Josef Glacier again.
In this case the changes in the balance profile due to a uniform change in air temperature are largest in the middle of the glacier. This is a consequence of the precipitation maximum found here. This result is in contrast with modelling results for most other glaciers, where changes in the specific balance tend to increase towards the glacier front (*figure 4.10*).
A complete coupling between mass balance and ice flow models is not necessary to deal with balance profiles that depend on the climatic state [by complete coupling we mean a synchronous integration of all relevant model equations]. Computed balance profiles like the ones shown in *figure 9.6* can be expressed in terms of polynomials or other functions (with altitude as variable) with coefficients that depend on air temperature. These are then easily incorporated in the forcing of the ice flow model.
Figure 9.7 shows the result of an inverse modelling exercise for Franz-Josef Glacier. The dependence of balance profiles on the temperature perturbation T' was included, as described above. Decadal temperature perturbations were determined by minimising D_L. Apparently, a temperature increase of about 0.7 K is sufficient to explain the rapid retreat of Franz-Josef Glacier during this century. Of course this does not prove that fluctuations in precipitation were unimportant. In fact, the same exercise can be performed on the assumption that only precipitation or cloudiness varied.

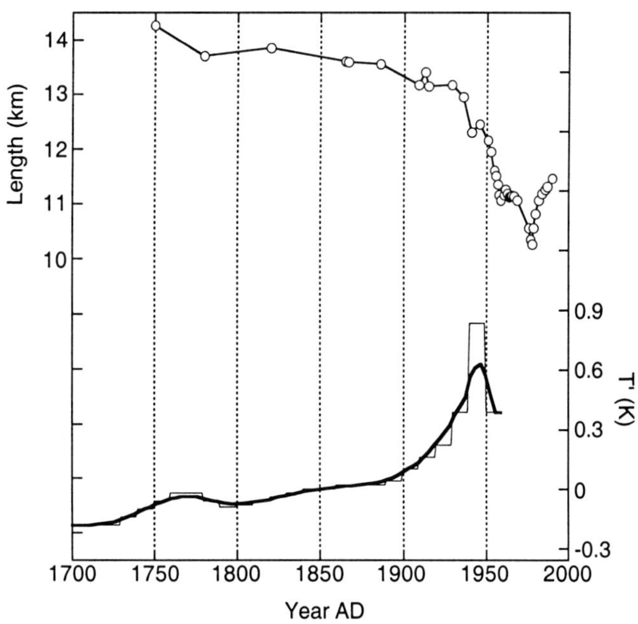

Figure 9.7. The length record of Franz-Josef Glacier (upper curve) and the annual mean temperature perturbation as reconstructed by inverse modelling (lower curve).

9.5 Climate noise in glacier records

Mass-balance observations have shown that the year-to-year variability is large (*figure 9.8*). Fortunately, glaciers tend to integrate the short-term variability and respond on a time scale that is interesting for climate studies. However, part of the longer-term fluctuations may very well be generated by the interannual noise in the specific balance. Glaciers can be considered as Brownian particles that are pushed around by never-ending irregular fluctuations of the equilibrium-line altitude. Superimposed on this we then have the "true" climate signal.

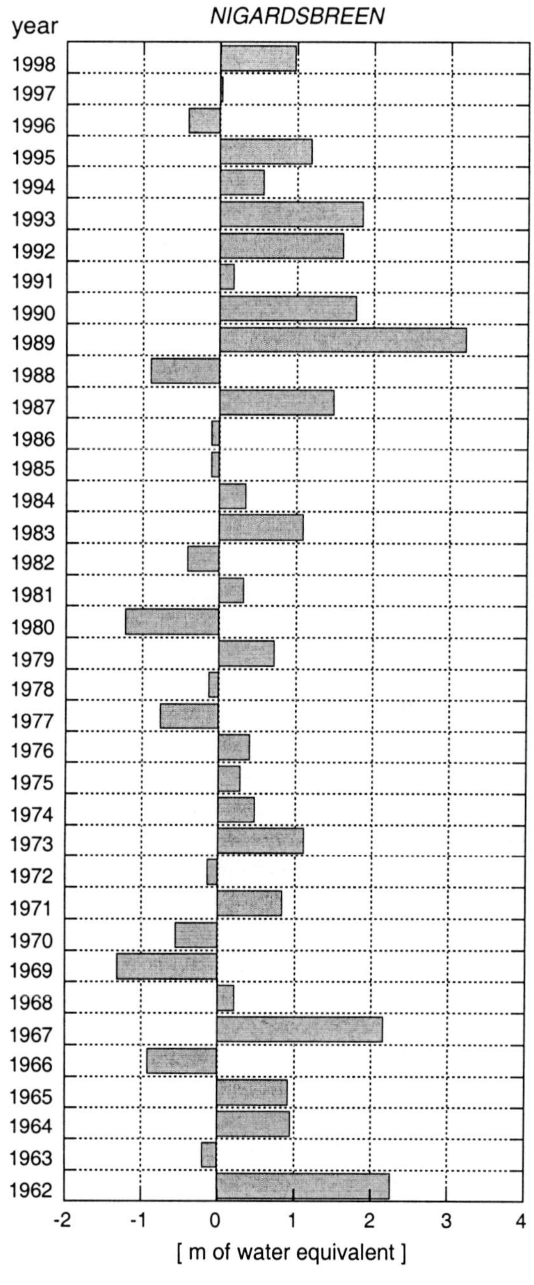

Figure 9.8. Observed specific balance of Nigardsbreen, Norway. Data from the World Glacier Monitoring Service with some additions from the Norges vassdrags- og energidirektorat, Oslo.

Considerations of this type were put in a more general framework ('stochastic climate models') by Hasselmann [11]. He argued that many quantities with which we characterise the climatic state owe part of their long-term variability to white noise forcing. One of the implications is that natural variability exists on a larger time-scale not directly related to external forcing or nonlinear dynamics [12].

In this section we investigate the implications of this type of behaviour for the interpretation of glacier records. We begin by referring again to the linear model, as described by eq. (9.2). The Fourier-transformed equation reads (where ω is the frequency):

$$i \omega \widetilde{L'} = - \frac{c \omega \widetilde{E'} + \omega \widetilde{L'}}{t_{rL}} \quad . \qquad (9.18)$$

The power spectrum of L', denoted by S_L, is then found by squaring this equation [13]:

$$S_L(\omega) = \frac{(c/t_{rL})^2 S_E(\omega)}{(1/t_{rL})^2 + \omega^2} \quad . \qquad (9.19)$$

Here S_E is the power spectrum of the forcing (fluctuations of the equilibrium-line altitude). If the forcing is white (year-to-year fluctuations in E' are not correlated) S_E is a constant. The power spectrum of glacier length is then a red spectrum, in which the variance decreases as ω^{-2}.

Figure 9.9 shows an example. A random generator was used to make a 1000-year record of equilibrium-line altitude with a standard deviation of 75 m. For realistic values for the glacier sensitivity and response time, the length variations are of considerable magnitude (typically 1 km on a century time scale).

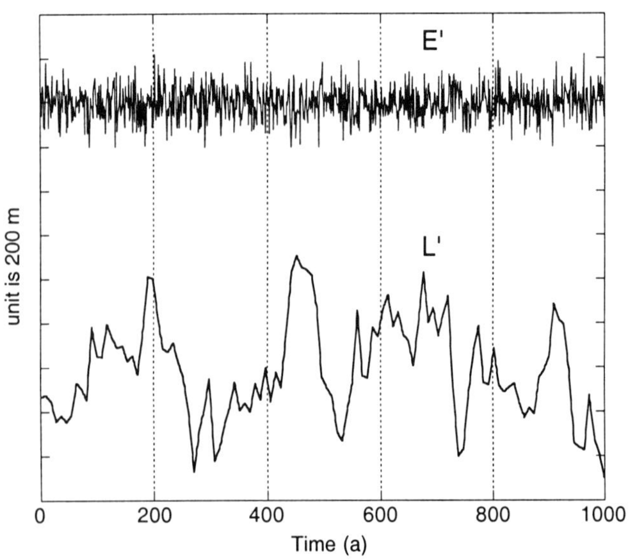

Figure 9.9. A synthetic glacier record obtained by integrating eq. (9.2) with white-noise forcing (E' has a standard deviation of 75 m). Model parameters: $c = 35$, $t_{rL} = 50$ a.

The total variance of L' can be obtained by integrating the power spectrum over frequency [14]. This yields the following relation between the standard deviation of glacier length (σ_L) and equilibrium-line altitude (σ_E):

$$\sigma_L = \left[\frac{\pi c^2}{\tau} \right]^{1/2} \sigma_E \ , \text{ where } \tau = \frac{t_{rL}}{\Delta t} \quad . \qquad (9.20)$$

In eq. (9.20) Δt is the time scale of the white-noise forcing (normally 1 year when we refer to fluctuations in equilibrium-line altitude or specific balance). Taking as characteristic values $c = 35$, $\tau = 50$ and $\sigma_E = 75$ m, we find $\sigma_L = 658$ m. This is an appreciable value and makes it clear that part of the variability of glacier length on the decadal to century time scale can be attributed to short-term weather fluctuations.

A linear model has its limitations and it is instructive to apply random forcing to the numerical flowline models discussed earlier. The forcing then takes the form

$$B(z,t) = B_{ref}(z) + B'(t) \quad . \qquad (9.21)$$

Here B'(t) is the annual balance perturbation with a white spectrum. The standard deviation of the annual balance perturbations, denoted by σ_B, can be estimated from long series of mass-balance observations. Some of these are listed in *table 9.1*. Values for σ_B range from about 0.25 (dry polar glaciers) to about 1 m a^{-1} (maritime glaciers). The value for Devon Ice Cap is relatively large (when compared to that for White Glacier), but this is mainly due to a significant positive trend in the mass balance data. This trend makes a 50% contribution to the value of σ_B. It also illustrates the fact that mass-balance fluctuations are not always purely white noise, because this would imply that the year-to-year perturbations are uncorrelated.

glacier	n	σ_B (m a^{-1})
Nigardsbreen (Norway)	36	1.06
Storbreen (Norway)	47	0.60
Storglaciären (Sweden)	50	0.75
Griesgletscher (Switzerland)	34	0.75
Hintereisferner (Austria)	43	0.54
Glac. de Sarennes (France)	47	0.90
Abramov Glacier (Kirghizstan)	23	0.73
Tuyuksu Glacier (Kazakhstan)	39	0.49
Peyto Glacier (Canada)	28	0.55
Devon Ice Cap (Canada)	34	0.55
White Glacier (Canada)	33	0.26

Table 9.1. Standard deviation of the mean specific balance as measured on a selection of glaciers. The first column gives the number of years used in the calculation of σ_B. Data from the World Glacier Monitoring Service (Zürich).

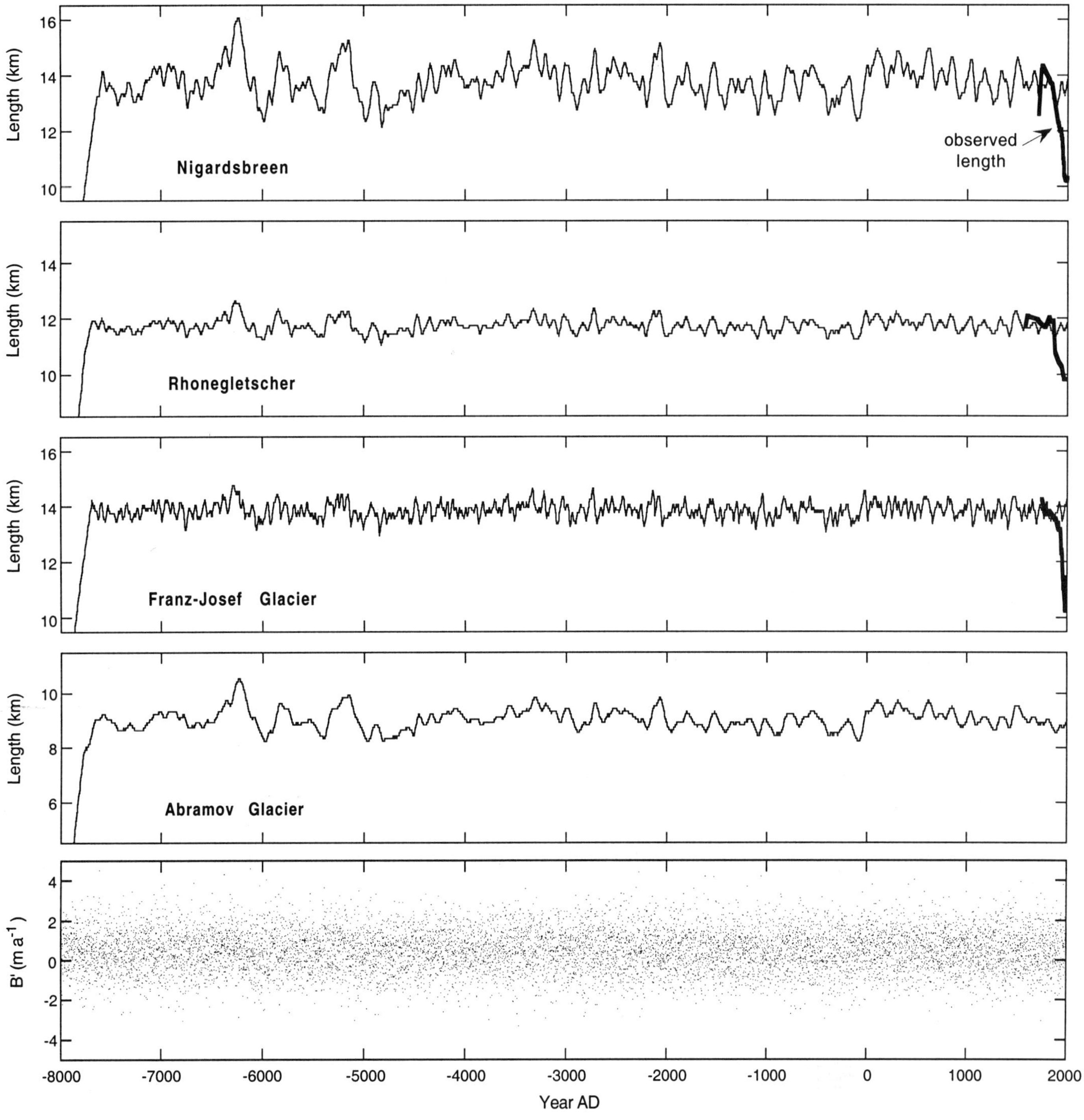

Figure 9.10. The calculated response of four glaciers to white-noise forcing for a 10 ka period. The lower panel shows B' for Nigardsbreen and Franz-Josef Glacier; for the Rhonegletscher and Abramov Glacier the same random data series was used, but scaled to give a value of 0.75 m a^{-1} for σ_B. The heavily drawn line curves show the observed length record. To facilitate a comparison, the vertical scales for the three glaciers are the same.

With regard to forcing of the flowline models the precise shape of the spectrum of the mass balance fluctuations is not important as long as the energy is contained in scales that are significantly smaller than the response time.

Figure 9.10 shows the modelled response of Nigardsbreen, the Rhonegletscher, Franz-Josef Glacier and Abramov Glacier. All integrations were started with no ice. The white-noise forcing for Nigardsbreen and Franz-Josef Glacier is shown at the bottom (σ_B = 1.05 m a^{-1}). For the Rhonegletscher and Abramov Glacier the same forcing was used, but scaled to give σ_B = 0.75 m a^{-1}.

The response of the glaciers is quite different. Nigardsbreen and Abramov Glacier show the largest fluctuations. Due to the smaller sensitivity and short response time of Franz-Josef

Glacier, the curve for this glacier has a different character, whereas the Rhonegletscher is in between.
The upper curve once more illustrates the extreme sensitivity of Nigardsbreen. The retreat during the last two centuries is impressive, but not larger than the retreat initiated by random forcing around the year -6250! In the case of the other glaciers the significance of the recent retreat seems to be larger.

9.6 Simulating glacier fluctuations from climate records

It is a simple fact that long, reliable and homogeneous series of meteorological quantities spanning at least two centuries do not exist. As a consequence, it is impossible to test the validity of glacier models for climate research, because this requires an accurately known forcing function for a period that is several times longer than the response time. It is only for small glaciers that model validation with climate forcing can be performed, but then a simple description of the mechanics as used in this book does not apply. On the other hand, because there are very few homogeneous meteorological series, a correct interpretation of documented glacier fluctuations is very valuable for research into climate change.
In many books and papers discussing climatic change over the past few centuries it has been stated, at least in a qualitative sense, that there is good agreement between proxy climate indicators, instrumental meteorological records and glacier fluctuations (for a general discussion see the books by Lamb [15], Grove [16], and Bradley and Jones [17]).
However, when it comes to explicit simulations of historical glacier variations with numerical models, things appear to be more complicated. With regard to the Alps, it is not too difficult to reproduce the glacier retreat since the middle of the 19th century (as was for instance demonstrated for Hintereisferner by Greuell [18]), but the main problem concerns the combined advance towards the maximum stand and the subsequent retreat (Glacier d' Argentière: [9], Rhonegletscher: [7]).
Tree rings have been put forward as a reliable climate proxy [17]. However, the problems encountered when one is interested in longer time scales are comparable to those with regard to glaciers. It is not clear that correlations between ring width and weather variables on an annual or decadal time scale hold on the century time scale. In the literature it has been suggested more than once that tree-ring width and equilibrium-line altitude are well correlated, because these parameters are both sensitive to summer temperature. It is therefore interesting to use tree-ring width as forcing for a glacier model, because this should give a consistent picture.
A well-known long tree-ring record is from Torneträsk, northern Sweden [19]. The ring width and latewood density of 65 trees were analysed to produce a record of 'summer temperatures in Fennoscandia' for almost 1500 years. It is tempting to use this series to force the Nigardsbreen model. Here the summer temperature is defined as the mean for the period April to August. From the SSC of Nigardsbreen (*figure 4.14*) it follows that the sensitivity of the balance rate to this parameter is about -0.5 m a^{-1} K^{-1}. Using this value to convert tree-ring summer temperature to mass-balance forcing then yields the upper curve in *figure 9.11*. The resulting glacier length is plotted together with the observed length record. Clearly, there is little correlation between the simulated and the observed glacier length. The retreat of Nigardsbreen is not simulated at all.
There may be several reasons for the discrepancy. Firstly, the summer temperature series derived from tree rings may not be valid on the time scale at which the glacier responds.

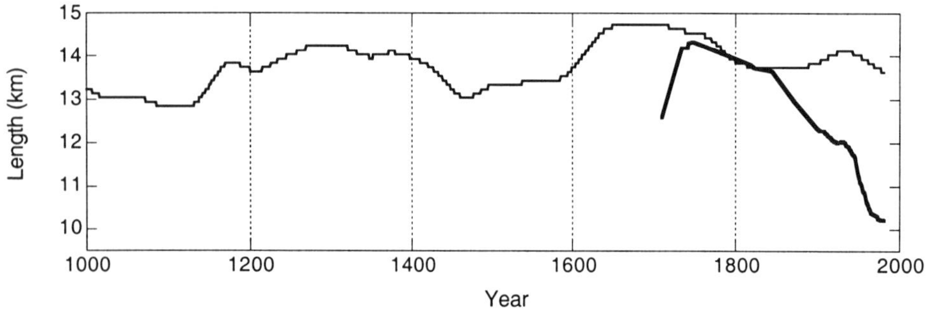

Figure 9.11. Results of a run with the Nigardsbreen model in which a tree-ring series from the Torneträsk region in northern Scandinavia was used as forcing (upper panel). The heavily drawn line is the observed length record.

Secondly, it may not be representative for a larger area, even for longer time scales. Thirdly, summer temperature was not the main factor determining mass-balance variations. And then, finally, the glacier model may be wrong. The discrepancy is hard to explain.
An earlier attempt to force a model for the Rhonegletscher with three rings from central Europe ('the Spessart series') also failed [7]. Nevertheless, it is of great importance to get climate proxies that agree, otherwise the credibility of paleoclimatic reconstructions will remain low.
The most successful simulation of a long historical glacier record was made by Schmeits [6]. In his study of the Unterer Grindelwaldgletscher he used long records of meteorological stations from sites close to the glacier and extended these backwards in time with the climate reconstructions for Switzerland made by Pfister [20]; the latter are based mainly on historical documents. The observed and the simulated records are plotted in *figure 9.12*.

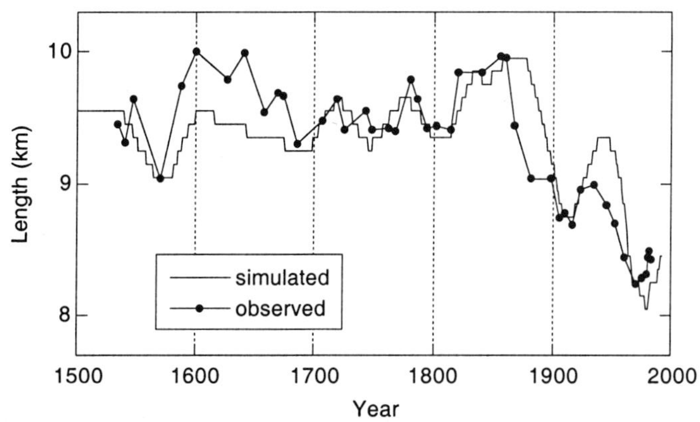

Figure 9.12. Observed and simulated length record of the Unterer Grindelwaldgletscher [6].

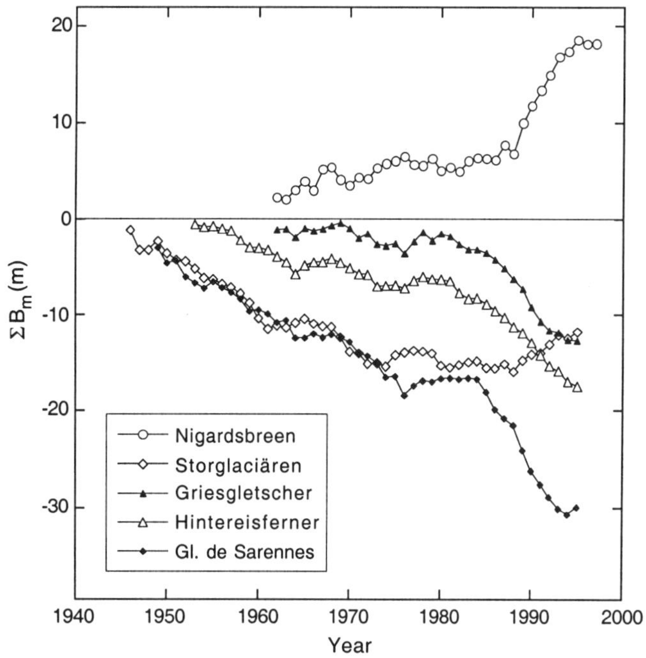

Figure 9.13. Cumulative mean specific balance for a selection of glaciers. Data from the World Glacier Monitoring Service.

Figure 9.14 shows the response of the Nigardsbreen model to such a forcing. It is an almost trivial response, but shows clearly that B_m and B' start to deviate as soon as the glacier geometry changes, i.e. after a few decades. When the glacier approaches a new steady state B_m of course goes to zero.
A step change in the climate forcing may be an extreme scenario, but it clearly illustrates the danger of using B_m as a climate indicator. The field glaciologist measuring B_m from year 140 to 160 would conclude that the climate is getting colder, but the signal is due entirely to the dynamic adjustment of the glacier.

9.7 Specific balance as a climate indicator?

The problem of using glacier length as an indicator of climate change is that it takes decades before the signal shows up. In contrast, the mean specific balance is a quantity that reflects a year's weather directly. It is general practice to characterise the climatic evolution on a decadal time scale by plotting the cumulative mean specific balance (ΣB_m) as a function of time (examples are shown in *figure 9.13*). But does B_m or ΣB_m really reflect a climatic signal ?
There can be little doubt that the short-term fluctuations in B_m are related to meteorological quantities. However, for longer time scales this is not so clear. Part of a long-term signal in B_m will be related to changes in glacier geometry [21]. This is best illustrated by analysing the response of a glacier in steady state to a sudden change in the balance profile, i.e. by imposing a step function for B'.

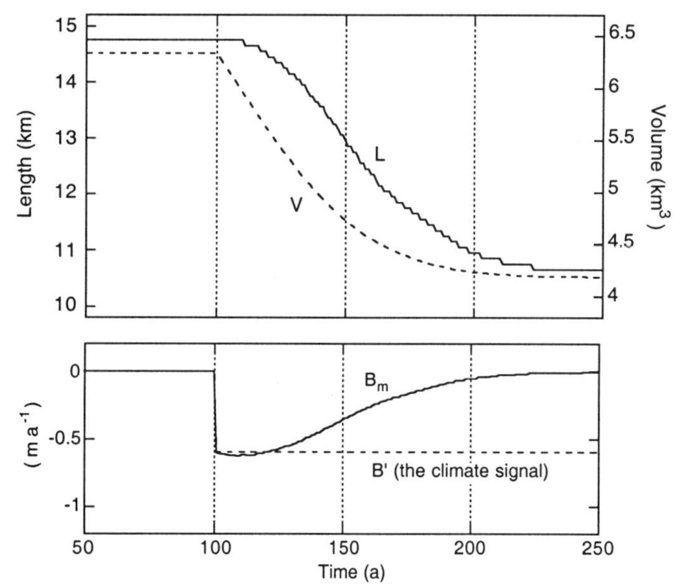

Figure 9.14. Computed response of Nigardsbreen to a sudden uniform change in the balance profile (B').

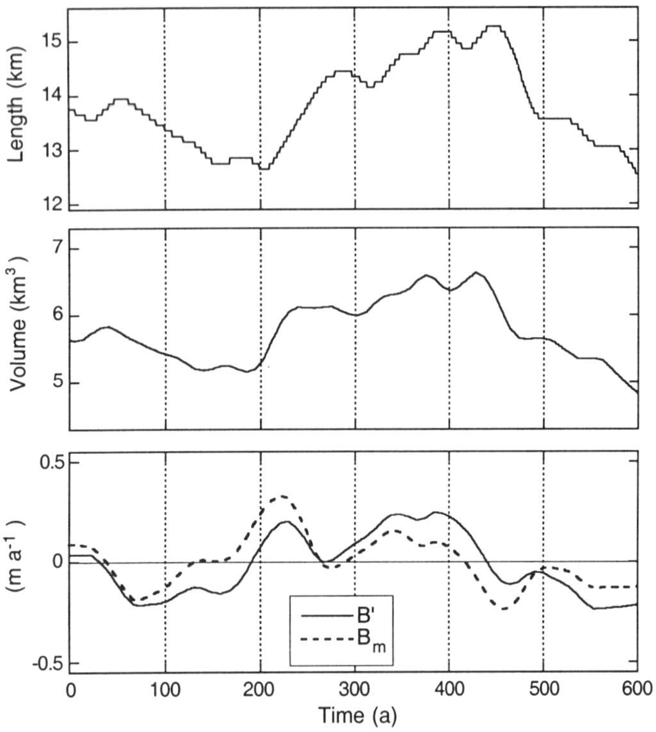

Figure 9.15. Analysis of a section of the 10 000-year run for Nigardsbreen (*figure 9.10*).

The point can be studied further by comparing B' and B_m from longer runs with a glacier model. *Figure 9.15* shows a 600-year close-up from the 10 000-year run for Nigardsbreen (*figure 9.10*). Because the forcing in this run was white noise, some smoothing has to be done to show the long-term signal clearly. First of all we note that glacier length lags ice volume by typically two decades, which is in agreement with the analysis of response times discussed earlier. Then it is clear that the B' and B_m curves are correlated, but not identical. The agreement depends very much on the state of balance of the glacier. Sometimes the curves are close, as in the first 80 years of the period shown; sometimes they diverge or have opposite signs. All in all, one should be very cautious about using B_m as a climate indicator. What B_m does tell us is if a glacier is losing or gaining mass. It is therefore an important quantity for judging the state of balance of a glacier.

9.8 Glacier models and climate models

Climate models play an important role in our understanding of past climatic change and they are being used more and more to make predictions about the future. However, today even the most sophisticated models have a spatial resolution that is not adequate to resolve glaciers and small ice caps. Temperature and precipitation fields are normally not accurate enough to be useful as input for a glacier mass-balance model. Therefore, to study the effect of climate change as generated by a climate model, it is common practice to look at relative changes. The simple idea behind this is that, even when meteorological quantities at a certain place are not accurately simulated, the calculated change in these quantities is basically correct. We will not discuss the validity of this assumption here, but we should not take it for granted because it is widely used!

With regard to the variability of glacier mass balance, changes in air temperature and precipitation (and possibly cloudiness) are the most important factors. The seasonal sensitivity characteristic (SSC, see section 4.8) is a convenient tool for calculating changes in the specific balance of a glacier from climate model output [see eq. (4.24)]:

$$\Delta B_m = \sum_{k=1}^{12} \left[SSC_{1,k} \left(T_j - T_{ref,k}\right) + SSC_{2,k} \left(P_j / P_{ref,k} - 1\right) \right] \quad (9.22)$$

As an illustration we consider the calculation of glacier response to output from a 10 000-year run with a climate model of intermediate complexity. This model has been designed by the Royal Netherlands Meteorological Institute and is called ECBilt [22]. It is a coupled model in which a full ocean model interacts with a relatively simple atmospheric model. The atmospheric part has richer dynamics than the more commonly used energy-balance climate models with parameterised dynamics [23]. An explicit hydrological cycle is simulated, and year-to-year variability is at an acceptable level. The spatial resolution of the model is poor (~500 km), but it can be integrated for long periods of time.

With ECBilt a global climate simulation has been carried out for the period 8000 BC to 2000 AD, i.e. for the last 10 000 years. In this run atmospheric composition has not been changed - only the insolation (due to the well-known changes in orbital parameters [24]). From this run, monthly mean values of precipitation and temperature can be extracted for any location and it is not difficult to calculate a mass-balance series for a certain glacier of which the SSC is known. Here we look at the results for Nigardsbreen (*figure 9.16*) and Abramov Glacier (*figure 9.17*). The contribution of changes in temperature and precipitation to the balance perturbation are shown separately. Again it should be realised that the calculated balance perturbations are defined with respect to the present-day glacier geometry.

For Nigardsbreen the temperature effect clearly dominates. The climate model predicts a modest cooling trend, particularly in summer, in response to the decreasing summer insolation in the high northern latitudes through the Holocene. Variations in precipitation contribute little, albeit there is a weak trend towards slightly drier conditions. The total mass balance perturbation increases by about 1 m a^{-1} during the 10 000-year period.

By taking the balance perturbation as an altitude-independent perturbation of the balance profile (eq. (9.21)), one can calculate the response of the Nigardsbreen model. The result is shown in the lower panel of *figure 9.16*.

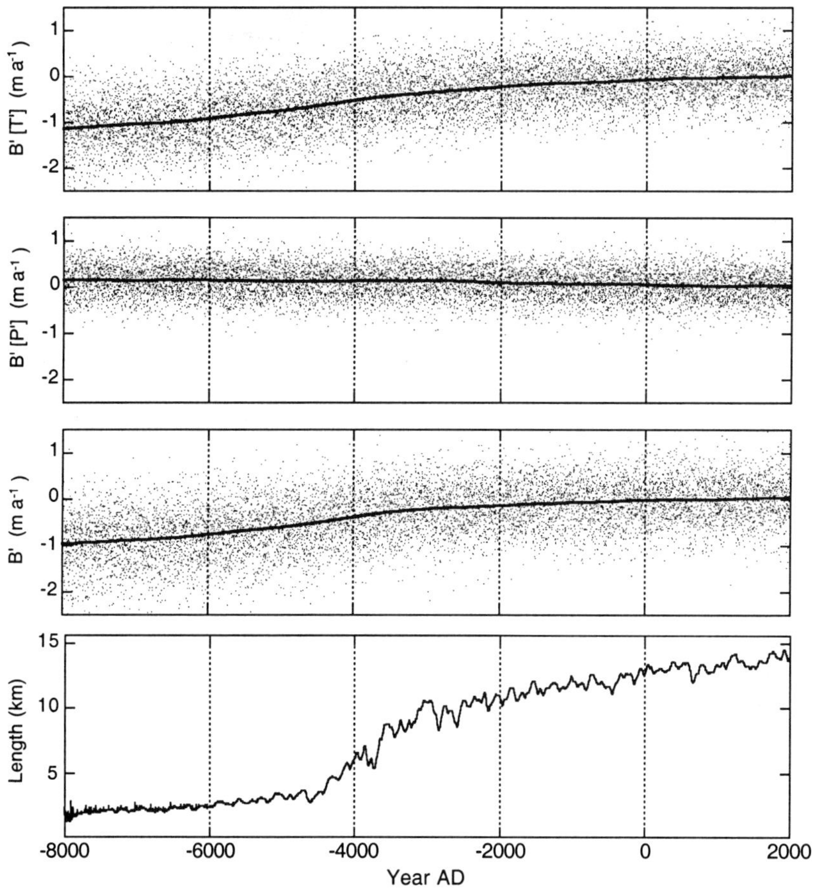

Figure 9.16. Holocene evolution of Nigardsbreen as simulated with a glacier model forced with output from a climate model (ECBilt, [22]). Shown are annual balance perturbation due to temperature change B'[T'], due to precipitation change B'[P'], the total balance perturbation B', and the resulting variations in glacier length.

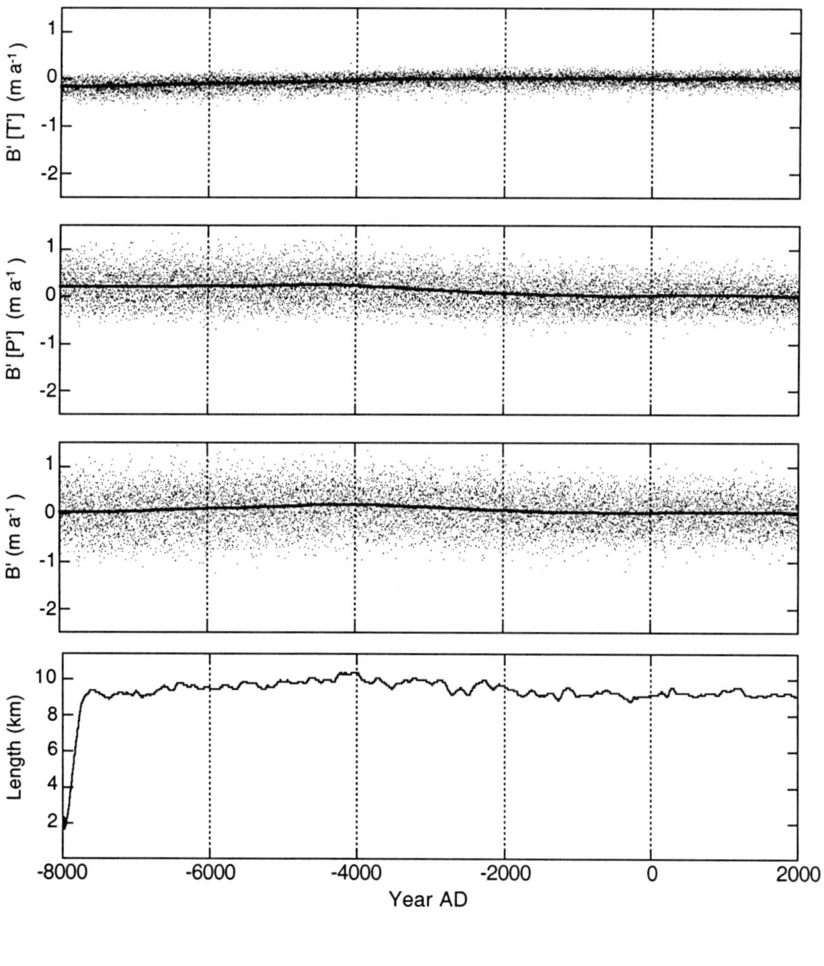

Figure 9.17. As *figure 9.16,* for Abramov Glacier.

The model suggests that a marked transition took place around 4000 BC. Around this time the glacier snout descended into the valley. Again, this reflects the strong nonlinearity of the sensitivity of Nigardsbreen to climate change. In the climate model no enhanced greenhouse forcing or any other external forcing (solar, volcanic) was included, so the model does not simulate the recent warming. As a consequence, the length of Nigardsbreen increases until the year 2000 AD. The short-term variability (decades to centuries) shown in *figure 9.16* is definitely less than was generated in the white-noise experiment (*figure 9.10*). This is because the year-to-year variability in the climate model is less than observed all over the globe. This is the price one has to pay when low resolution and simpler dynamics are used to keep processing times short. Nevertheless, this does not mean that the long-term trends are not well simulated.

For the location of Abramov Glacier the variability is even less, especially with regard to temperature. Temperature and precipitation decrease slightly during the Holocene (note that at the latitude of Abramov Glacier the orbital forcing is much smaller). The final result in terms of glacier length shows a remarkable constancy. However, as for Nigardsbreen the short-term variability is underestimated.

The results displayed in *figures 9.16/9.17* are not meant to represent a Holocene glacier history for Nigardsbreen and Abramov Glacier, but they merely serve to show the potential of forcing glacier models with climate models. With increasing computer power the quality and the potential of climate models are likely to increase, and the generation of paleorecords along the lines discussed here may become a powerful tool in paleoclimatology.

References and further reading

1. O. Sigurdsson (1998): Glacier variations in Iceland 1930-1995. From the database of the Icelandic Glaciological Society. *Jökull* **45**, 3-26.

2. W. Haeberli (1998): Historical evolution and operational aspects of worldwide glacier monitoring. In: *Into the Second Century of World Glacier Monitoring - Prospects and Strategies* (W. Haeberli, M. Hoelzle, and S. Suter, eds.). UNESCO publishing, Paris, 35-51.

3. J.F. Nye (1965): A numerical method of inferring the budget history of a glacier from its advance and retreat. *Journal of Glaciology* **5** (41), 589-607.

4. J. Oerlemans (1994): Quantifying global warming from the retreat of glaciers. *Science* **264**, 243-245.

5. W. Haeberli and M. Hoelzle (1995): Application of inventory data for estimating characteristics of and regional climate-change effects on mountain glaciers: a pilot study with the European Alps. *Annals of Glaciology* **21**, 206-212.

6. M.J. Schmeits and J. Oerlemans (1997): Simulation of the historical variations in length of the Unterer Grindelwaldgletscher. *Journal of Glaciology* **43** (143), 152-164.

7. A. Stroeven, R.S.W. van de Wal and J. Oerlemans (1989): Historical fluctuations of the Rhone glacier: simulation with a numerical model. In: *Glacier Fluctuations and Climatic Change.* (ed.: J. Oerlemans), Reidel, 391-405.

   J. Wallinga and R.S.W. van de Wal (1998): Sensitivity of the Rhone glacier to climate change: experiments with a one-dimensional flow-line model. *Journal of Glaciology* **44** (147), 383-393.

8. Ph. Huybrechts, P. de Nooze and H. Decleir (1989): Numerical modelling of Glacier d'Argentière and its historical front variations. In: *Glacier Fluctuations and Climatic Change.* (ed.: J. Oerlemans), Reidel, 373-389.

9. T. Jóhannesson, C.F. Raymond and E.D. Waddington (1989): Time-scale for adjustment of glaciers to changes in mass balance. *Journal of Glaciology* **35**, 355-369.

10. J. Oerlemans (1997): Climate Sensitivity of Franz-Josef Glacier, New Zealand, as revealed by numerical modelling. *Arctic and Alpine Research* **29** (2), 233-239.

11. K. Hasselmann (1976): Stochastic climate models. Part 1. Theory. *Tellus* **18**, 473-483.

12. S. Manabe and R.J. Stouffer (1996): Low frequency variability of surface air temperature in a 1000 yr integration of a coupled atmosphere-ocean-land surface model. *Journal of Climate* **9**, 376-393.

13. R.B. Blackman and J.W. Tukey (1958): The measurement of power spectra. Dover, New York, 190 pp.

14. T.T. Soong (1973): Random differential equations in science and engineering. Academic Press, New York, 327 pp.

15. H.H. Lamb (1977): *Climate, Present, Past and Future.* Methuen (London), 835 pp.

16. J.M. Grove (1988): *The Little Ice Age.* Methuen (London), 498 pp.

17. R.S. Bradley and P.D. Jones, editors (1995): *Climate Since A.D. 1500.* Routledge (London), 2nd ed., 706 pp.

18. J.W. Greuell (1992): Hintereisferner, Austria: mass-balance reconstruction and numerical modelling of the historical length variations. *Journal of Glaciology* **38**, 233-244.

19. K.R. Briffa, T.S. Bartholin, D. Eckstein, P.D. Jones, W. Karlén, F.H. Schweingruber and P. Zetterberg (1990): A 1,400-year tree-ring record of summer temperatures in Fennoscandia. *Nature* **346** (6283), 434-439.

K.R. Briffa, P.D. Jones, T.S. Bartholin, D. Eckstein, F.H. Schweingruber, W. Karlén, P. Zetterberg and M. Eronen (1992): Fennoscandian summers from AD 500: temperature changes on short and long time scales. *Climate Dynamics* **7**, 111-119.

20. C. Pfister (1984): Klimageschichte der Schweiz 1525-1860. *Academia Helvetica* **6** (1).

21. W. Haeberli (1994): Accelerated glacier and permafrost changes in the Alps. In: Beniston, M. (Ed.): *Mountain environments in changing climates.* Routledge, London, 91-107

22. J.D. Opsteegh, R.J. Haarsma, F.M. Selten and A. Kattenberg (1998): ECBILT: a dynamic alternative to mixed boundary conditions in ocean models. *Tellus* **50A**, 348-367.

S.L. Weber and J. Oerlemans (200x): Holocene glacier variability: three case studies using an intermediate-complexity climate model. *The Holocene*, in press.

23. R. Bintanja (1997): Sensitivity experiments performed with an energy balance atmosphere model coupled to an advection-diffusion ocean model. *Theoretical and Applied Climatology* **56**, 1-24.

24. A.L. Berger (1978): Long-term variations of caloric insolation resulting from the earth's orbital elements. *Quaternary Research* **9**, 139-167.

A.L. Berger (1978): Milankovitch theory and climate. *Review of Geophysics* **26**, 624-657.

10. The global picture and sea-level rise

10.1 Worldwide glacier retreat

As shown in *figure 9.1*, many glaciers have retreated in the last 100 to 200 years. To extract a climate signal, one needs to apply inverse modelling to these records and to many others as well. However, it is not simple to calculate the sensitivities and response times of all these glaciers and it will be some time before the goal is reached.

As an intermediate step one can ignore the problem of response time and look only at the longer-term signal. As discussed in Chapter 9, the author adopted this approach [1] and a brief description of this work is repeated here.

Formally, climate and glacier fluctuations over a specific period can be viewed as a linear trend on which smaller-scale fluctuations are imposed. One can try try to find the linear component, thereby assuming that glacier extent is in balance with climate. The longer the period under consideration, the more accurate this assumption will be.

The work was based on the data on glacier length stored at the World Glacier Monitoring Service (in 1993). By requiring records to have a first data point before 1940, a sample of only 48 glaciers was found. For the period 1850 - 1990 the average length of the 48 records of glacier length used turned out to be 94 years. The glaciers were: Ape, Athabasca, Asulkan, Clendenning, Drummond, Fleur d. Neiges, Fyles, Griffin, Havoc, Illecillewaet, New Moon, Noeick, Peyto, Purgatory, Saskatchewan, Sentinel, Staircase, Surf, Terrific, Tsoloss, Victoria, Wedgemount (Canada); Eel, White (U.S.A.); Samarin, Hans, Paierl (Spitsbergen); Vatnajökull (Iceland); Engabreen, Nigardsbreen (Norway); Rhonegletscher, Unterer Grindelwaldgletscher (Switzerland); Hintereisferner (Austria); Argentière (France); Bolshoy Abyl-Oyuk, Geblera, Bolshoy Maashey, Maliy Aktru, Rodzevicha, Sapozhnikova (Kazachstan); Chungpar-Tash., Chogo Lungma, Minapin (Pakistan); Cartensz, Meren (Irian Jaya); Lewis, Tyndal (Kenya); Franz Josef (New Zealand).

For all these glaciers the mean rates of retreat R were calculated from the data. All glaciers in the sample became shorter, the mean rates of retreat being between 1.3 and 86 m a^{-1} (*figure 10.1*).

Although for this study the glaciers were not modelled in any way, a scaling procedure was applied to deal with the fact that maritime glaciers are more sensitive to climate change. In the preceding chapters we found that glaciers with smaller slopes are more sensitive than those with steep slopes. These arguments were used to define a rate of retreat $\tilde{R}$ corrected for slope and climate sensitivity according to

$$\tilde{R} = \frac{s}{\bar{s}} \frac{\overline{C_T}}{C_T} R \quad . \qquad (10.1)$$

Here s is a characteristic slope and $\bar{s}$ its mean value over the sample of 48 glaciers. C_T is the climate sensitivity (change of mean balance with temperature as defined in section 4.9). Values for s were estimated from data in the World Glacier Inventory and C_T from the mass-balance modelling as discussed in section 4.9. $\tilde{R}$ turned out to have a significantly smaller variance than R. The effect is caused mainly by corrections to the very large retreat rates for the glaciers in Svalbard (these have small slopes) and the glaciers in Norway (these are maritime and consequently have large values for C_T). For global mean values, the signal-to-noise ratio, defined here as the mean rate of retreat divided by the standard deviation within the sample of 48 glaciers, increases from 1.22 for the non-scaled values to 1.93 for the scaled values.

When grouped into regions, the values of $\tilde{R}$ have the same order of magnitude (*figure 10.1*). A plot of all values of R and $\tilde{R}$ as a function of latitude also shows a rather small variation (*figure 10.2*). Can we conclude from this that the climate warming during the last 100 or 150 years has been rather uniform over the globe? Perhaps. The glacier signal, in any case, appears to be remarkably consistent.

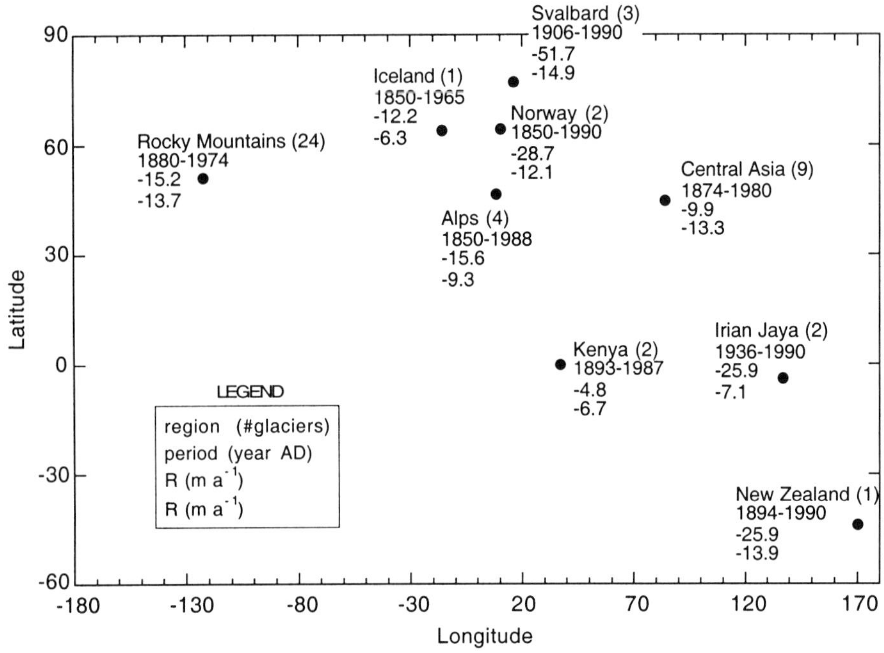

Figure 10.1. Results of an analysis of glacier length changes [2], summarised for nine regions. For each region are indicated: year of beginning and end of the records (mean values), the mean rate of retreat, the mean scaled rate of retreat

We now postulate that the glacier retreat has been caused by a uniform linear increase in atmospheric temperature. How large then must this warming have been? Since response times are not considered, one has to assume that glaciers were close to a steady state all the time. Values of dL/dT_{fa} (see section 8.2, *table 8.2*) can be used to relate a change in glacier length to a warming rate. Since modelled glaciers are not necessarily representative for the sample of 48 glaciers, the scaling procedure has to be applied again:

$$\widetilde{\frac{dL}{dT_{fa}}} = \frac{1}{N}\sum_{k=1}^{N} \frac{dL_k}{dT_{fa}} \frac{s_k}{\overline{s}} \frac{\overline{C_T}}{C_{T,k}} \quad . \qquad (10.2)$$

In this expression N is the number of modelled glaciers and k refers to the individual glaciers. The left-hand side therefore represents a characteristic value of dL/dT_{fa} that can be applied to the mean scaled retreat of the sample. On the basis of modelling done for Hintereisferner, Nigardsbreen, Rhonegletscher, Glacier d'Argentière and Lewis glacier a value of 2.0 km K^{-1} was found for $\widetilde{dL/dT_{fa}}$.

The average starting and finishing dates of all records are 1884 and 1978. Over this period the scaled mean glacier retreat is 1.23 km, implying a 0.62 ± 0.09 K warming over this period, corresponding to 0.66 ± 0.10 K per 100 years (the uncertainty is based on a 30% error in the individual estimates of slope and climate sensitivity and on the assumption that the regions, but not the glaciers within a region, have independent errors). The total error may be twice as large (of the order of 0.2 K). Factors other than temperature affect the size of glaciers. The response times for the largest glaciers in the sample may be larger than desired for this study. Some of the records begin in 1850, others in 1925. However, assigning a smaller weight to the cluster of records from the Northern Hemisphere mid-latitudes has no effect. Differentiating between records of different length does not change the result significantly either.

In *figure 10.3* the calculated temperature trend is compared with instrumental measurements. The agreement is good and we conclude that the worldwide retreat of glaciers provides an independent confirmation of the fact that the global climate has become warmer.

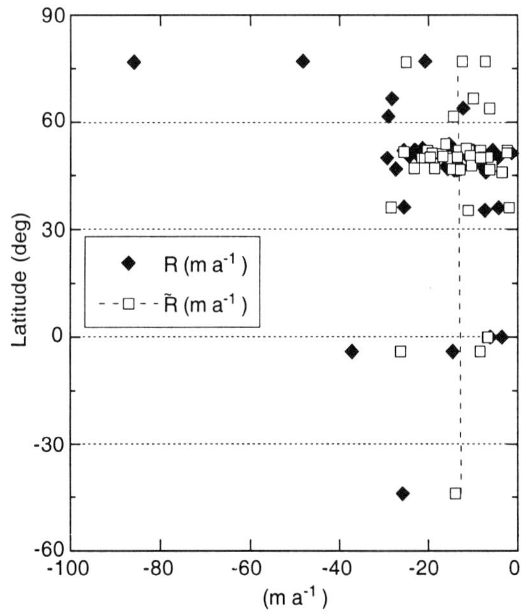

Figure 10.2. Mean rate of retreat, now plotted as a function of geographical latitude. The dashed line is a linear fit to the corrected retreat rates.

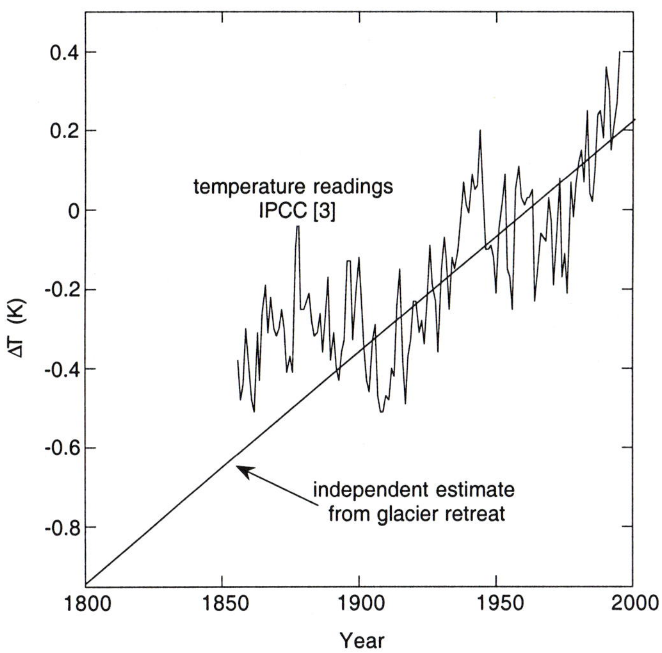

Figure 10.3. A comparison between instrumental temperature measurements (global mean) and the temperature trend needed to explain the retreat of glaciers.

10.2 Glaciers and sea-level change

Although this book is about valley glaciers, when the global picture is discussed one needs to give attention to other ice masses as well. One of the important aspects of glacier retreat is the potential contribution that it makes to sea-level rise. It is generally accepted that the global sea level is currently rising at a rate of 10 to 25 cm per century. The uncertainty is large, because tide-gauge measurements are very unevenly spread over the globe. At many locations local vertical movements are significant, making an interpretation in terms of global changes difficult. Soon, radar altimetry from satellites will give us a more accurate picture of what happens to the ocean surface, but it will be some time before a reliable estimate of global sea-level change over several decades becomes available.

Anyway, most investigators believe that thermal expansion of ocean water and changes in the volume of land ice are the most important processes with regard to sea-level change on a century time scale. It is an interesting question to what extent the retreat of glaciers has contributed to sea-level rise. Before discussing this, we need to put glaciers and small ice caps in perspective, and a few basic characteristics of the large ice sheets have to be considered.

The Antarctic Ice Sheet stores roughly 90% of the land ice on earth, the Greenland Ice Sheet about 9% and glaciers and small ice caps much less than 1% [3]. How, then, can glaciers be of any significance?

First of all, with regard to changes in the exchange of mass with the atmosphere, area is a more relevant quantity than volume. Glaciers and ice caps have a relatively large area compared to their volume. Secondly, the mean accumulation over glaciers and ice caps is much larger than over the dry polar ice sheets. The mean accumulation on the Antarctic Ice Sheet is estimated to be about 0.17 m a^{-1} , on the Greenland Ice Sheet 0.35 m a^{-1} [3]. In contrast, the mean accumulation on glaciers and ice caps has been estimated to be 1 m a^{-1} [4]. A turnover time-scale (t_{over}) can be defined as the volume of an ice mass divided by the accumulation rate. This yields the values listed in *table 10.1*. For glaciers and ice caps a value of 265 a is found, which is much smaller, of course, than the turnover time of the ice sheets. Perhaps this value should be a bit larger, say 500 a, because outside the polar regions a significant part of the precipitation on glaciers falls as rain and runs off.

A possible way of looking at the potential importance of ice mass for sea-level change is to calculate what would happen if accumulation were to be reduced by 10%, because this highlights the importance of glacier area for the initial response to rapid climate change (*figure 10.4*). Clearly, the Antarctic ice sheet would make the largest contribution, but the Greenland ice sheet and glaciers and ice caps would play a significant role as well. The numbers are certainly more alike than the differences in volume would at first sight suggest. The estimates shown in *figure 10.4* give only the initial response. After some time, depending on the response time of the ice mass, the geometry will start to feed back to the balance in order to restore a new equilibrium.

	area (10^6 km^2)	volume (10^6 km^3)	sea-level eq. (m)	precip (m a^{-1})	t_{over} (a)
Antarctic Ice Sheet	12.1	29	73	0.17	14098
Greenland Ice sheet	1.7	2.95	7.4	0.35	4956
glaciers and ice caps	0.68*	0.18**	0.5**	1**	265

Table 10.1. A few physical characteristics of ice sheets and glaciers. Precision normally within 10%, but 15% for * and 20% for **

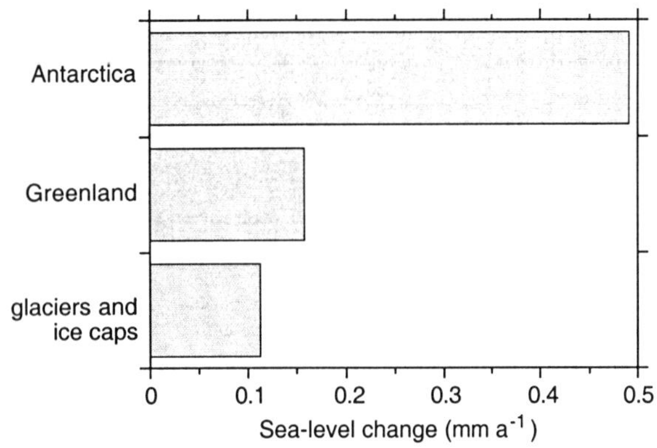

Figure 10.4. The initial response of global mean sea level to a 10% reduction in the accumulation on ice sheets, glaciers and ice caps.

In the light of the anticipated greenhouse warming the response to temperature rise may be the most interesting aspect to look at. As a start we consider the static approach, which implies that changes in the geometry of ice masses are not taken into account. First of all a reference state has to be chosen, assigned to a reference period. For instance, we could take the period 1961-1990 as the climatological norm, and use the mean specific balance and glacier area to define the reference state. In the static approach, the change in volume of glacier k with time due to climate change can be described as

$$\frac{dV_k}{dt} = A_k\left(B_{0,k} + B_k'\right) \quad . \tag{10.3}$$

Here V_k is glacier volume, A_k glacier area, $B_{0,k}$ the mean specific balance over the reference period and B_k' the balance perturbation associated with a change in climate (with respect to the climate in the reference period). It is clear that eq. (10.3) can only be valid for sufficiently small perturbations and over sufficiently small time intervals, because otherwise the assumption of constant glacier area is not justified (in fact, we have assumed glaciers to be infinite sources of ice).

Can we use this approach to study the potential effect of a uniform global warming of 1 K? To arrive at an order-of-magnitude estimate we set $B_{0,k} = 0$ and sum over all glaciers in the world:

$$\frac{dV_{tot}}{dt} = \sum_{\text{all k}} A_k B_k' \quad . \tag{10.4}$$

Since there are tens of thousands of glaciers, it is not so easy to evaluate eq. (10.4) for a 1 K warming. In an earlier study an attempt was made to achieve this by dividing glaciers into groups and by assigning characteristic values of B_k' using the relation between sensitivity and the precipitation regime [5, 4]. All glaciers were grouped according to geographical location, and precipitation was estimated mainly from climatological maps. Here we repeat the exercise, but with the glaciers and ice caps surrrounding the main Greenland Ice Sheet as an additional group. The groups of glaciers are listed in the table below *(table 10.2)*, for the spatial distribution see *figure 2.1*. Virtually all regions with glaciers are listed here, but in some cases the ice-covered area is very small (at least this holds for all tropical glaciers).

Number	Region	Latitude	Longitude	Area (1000 km^2)	Precip (m a^{-1})
1	Ellesmere	80°N	80°W	80	0.22
2	Axel Heiberg	79°N	92°W	11.7	0.22
3	Devon island	75°N	82°W	16.2	0.225
4	N-Baffin island	72°N	78°W	18.5	0.312
5	S-Baffin island	67°N	66°W	18.5	0.5
6	Bylot island	73°N	78°W	5	0.22
7	Brooks range	68°N	150°W	0.722	0.375
8	Alaska range	63°N	151°W	13.9	0.875
9	Talkeetna mts.	62°N	149°W	0.8	0.875
10	Kilbuck mts.	61°N	160°W	0.23	0.75
11	Aleutian islands	54°N	166°W	0.96	1.5
12	Alaska peninsula	57°N	157°W	1.25	1.25
13	Seward peninsula	60°N	149°W	1	3
14	Kenai mts.	60°N	150°W	4.6	1.25
15	Wrangell mts.	62°N	143°W	8.3	1.25
16	Chugach mts.	61°N	145°W	21.6	1.75

Number	Region	Latitude	Longitude	Area (1000 km^2)	Precip (m a^{-1})
17	St.Elias mts.	61°N	141°W	11.8	1.5
18	Coast mts.	59°N	134°W	10.5	2.5
19	Stikine river	57°N	131°W	5.07	1.5
20	Yukon territory	61°N	140°W	10	1.5
21	Selwyn mts.	62.5°N	128°W	0.6	0.75
22	Rocky mts.	58°N	125°W	0.3	0.75
23	Coast mts.	51°N	125°W	13.5	2.5
24	Olympic mts.	48°N	123°W	0.046	4
25	N-Cascades	48.3°N	121°W	0.266	2.5
26	Mt.Raineer / Adams	46°N	122°W	0.063	4
27	Rocky mts.	52°N	118°W	3	1
28	Selkirk mts.	51°N	117°W	1	1.5
29	Rocky mts.	44°N	109°W	0.074	0.562
30	Sierra Nevada	37°N	118°W	0.051	1
31	Mexico	19°N	97°W	0.004	2.5
32	Mexico	19°N	98.5°W	0.004	1.25
33	Mexico	19°N	100°W	0.003	1.75
34	S.Nev. de Merida	8.5°N	71°W	0.003	1.5
35	Nev. del Huila	3°N	76°W	0.019	1.5
36	Nev. del Ruizy/ Tolima	4.7°N	75.4°W	0.024	3
37	Nev. del Cocuy	6.5°N	72.3°W	0.039	1.25
38	S.Nev.de Santa Marta	10.8°N	73.6°W	0.019	4
39	Ecuador	0.8°S	78.2°W	0.111	3.5
40	N. Cordilleras	10°S	77°W	0.873	1.5
41	N. Cordilleras	12°S	75.5°W	0.258	1
42	S. Cordilleras	15°S	73°W	0.649	0.813
43	Cordilleras Oriental	15.6°S	68°W	0.51	1.12
44	Andes, 30-16°S	23°S	67.5°W	2.8	0.375
45	Chile	34°S	70°W	0.743	1.5
46	Mendoza/San Juan	33°S	70°W	1.17	1.5
47	Neuquen	37°S	70°W	0.156	3.5
48	Chile	43°S	72°W	3.57	3
49	N. Patagonia	47°S	73.5°W	5.72	2
50	S. Patagonia	50°S	73.5°W	16.9	2
51	Tierra del Fuego	54.5°S	70.5°W	2.7	2
52	S.Atlantic islands	57°S	35°W	4.1	1
53	Vatnajokull	65.5°N	17°W	8.3	4
54	Langjokull	65°N	20°W	0.953	4
55	Hofsjokull	65.5°N	19°W	0.925	2.5
56	Myrdalsjokull	63.5°N	19°W	0.6	4
57	Drangajokull	66°N	22°W	0.16	2.5
58	Svalbard	79°N	18°E	36.6	0.375
59	Franz Josef Land	81°N	57°E	13.7	0.312
60	Novaya Zemlya	75°N	57°E	23.6	0.5
61	Severnaya Zemlya	79°N	98°E	18.3	0.25
62	N. Scandinavia	67.5°N	17°E	1.08	2.5

Number	Region	Latitude	Longitude	Area (1000 km^2)	Precip (m a^{-1})
63	S. Norway	61.7°N	7.4°E	1.63	4
64	Alps	46°N	7.5°E	2.91	2
65	Pyrenees	42.5°N	0°E	0.012	2
66	Bol'shoy Kavkaz	42.5°N	44°E	1.34	1.5
67	Turkey	38°N	43.5°E	0.024	1.25
68	Iran	38°N	46.5°E	0.02	0.75
69	Severniy Ural	66°N	61°E	0.8	1
70	Ili+Chu	43°N	77.5°E	2.3	1
71	Syrdar'ya / Amudar'ya	38°N	71°E	10.7	1
72	Tarim Ba	42°N	77.5°E	3.32	0.75
73	Pakistan / India	36°N	76°E	40	1.75
74	Nanga Parbat	35.3°N	74.6°E	0.301	1.25
75	Chitral	36.4°N	72.4°E	1.35	2
76	Himalaya / Nepal	28°N	87°E	12.7	0.75
77	Bhutan	28°N	90°E	1.34	0.75
78	Niangtingtanggula	30°N	89°E	7.54	1
79	Qilian Shan	39°N	99°E	2.38	0.75
80	Altai Shan	47°N	91°E	0.293	0.375
81	Kunlun Shan	36°N	87°E	11.6	0.25
82	Tyan Shan	42°N	80°E	9.55	1
83	Tanggula Shan	33°N	91°E	2.08	0.75
84	Hengduan Shan	28°N	99°E	1.46	1
85	Gangdisi Shan	31°N	82°E	2.19	0.625
86	Dzungarsky Alatau	45°N	80°E	0.91	1
87	Qiangtang Plateau	34°N	87°E	3.19	0.25
88	Orulgan	68°N	128°E	0.018	0.625
89	Cherskogo	65°N	145°E	0.372	0.875
90	Koryakskoye Khrebet	62.5°N	171°E	0.26	1.25
91	Suntar Khayata	62.5°N	141°E	0.202	1
92	Kodar	57.5°N	117°E	0.019	0.75
93	Kamchatka	57.5°N	160°E	0.874	0.5
94	Mt. Kenya	0.2°S	37.3°E	0.001	1.75
95	Kilimandjaro	3.1°S	37.3°E	0.005	1.37
96	Ruwenzori	0.4°N	30°E	0.005	1.37
97	Irian Jaya	4°S	137°E	0.007	3.75
98	New Zealand	43.5°S	170°E	1	1.87
99	Kerguelen island	49°S	69°E	0.45	1.07
100	Heard island	53°S	73.5°E	0.45	1.72
101	Greenland (glaciers)	78°N	32°W	150	0.35
	TOTAL			677.9	
	World ocean			362000	

Table 10.2. A regional partition of all glaciers and ice caps outside the main ice sheets of Greenland and Antarctica. The last column gives the characteristic precipitation for the region in which the glaciers and ice caps are located. All glaciers and ice caps on Greenland that are not connected to the main ice sheet are lumped together under number 101.

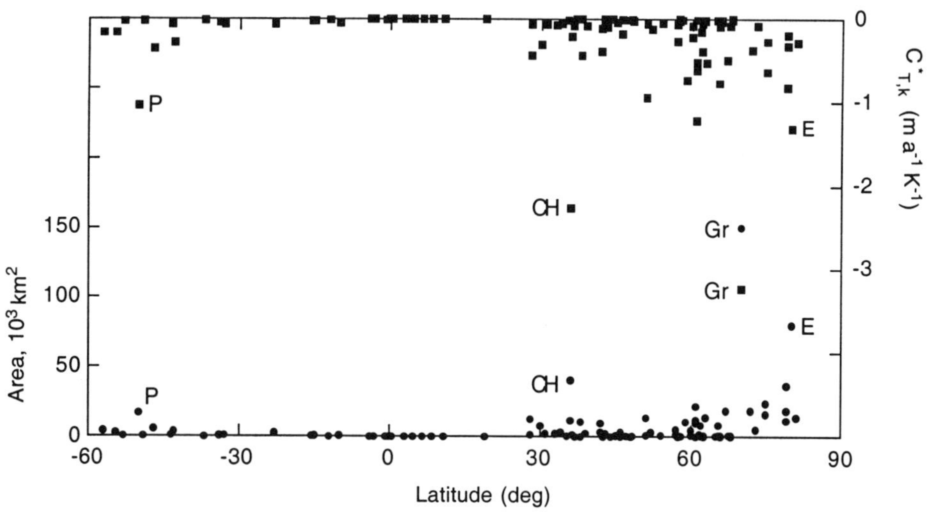

Figure 10.5. The glacier regions of *table 10.2* plotted against latitude. The lower part of the figure shows the areas, the upper part shows the area-weighted sensitivity as defined in the text. A few ice masses are labelled: glaciers and ice caps of Greenland (Gr), Ellesmere (E), central Himalaya (CH) and the southern Patagonian ice field (P).

The global mean sensitivity to temperature change (uniform through the year) can be defined as

$$C_{T,glm} = \frac{1}{A_{tot}} \sum_{k=1}^{101} A_k \, C_{T,k} \; . \tag{10.5}$$

Here A_{tot} is the total glacier area and $C_{T,k}$ the sensitivity as defined in section 4.7. The values for the individual groups of glaciers can be estimated from eq. (4.20):

$$C_{T,k} = -0.271 \left(P_{ann,k}\right)^{0.597} \; . \tag{10.6}$$

For most regions the value of $P_{ann,k}$ was taken from climatological maps. Except for the flat ice caps of the Arctic, precipitation was corrected for the fact that glaciers form in the wettest places. The correction applied, +25%, is just a guess, however. The value of $C_{T,glm}$ thus found is -0.22 m a^{-1} K^{-1}. This may seem rather small. It reflects the fact that the arctic ice caps in a relatively dry climate dominate the picture. This observation also implies that the result has to be considered with some caution, because most mass balance studies and related modelling work were done for glaciers in the mid-latitudes. More glacio-meteorological studies on the arctic ice caps are certainly needed.

To visualise the relative importance of the regions with regard to sea-level change one can weight the sensitivity with the area:

$$C^*_{T,k} = \frac{101 \, A_k}{A_{tot}} \, C_{T,k} \; . \tag{10.7}$$

The values for $C^*_{T,k}$ are also plotted in *figure 10.5*. It can be seen that some of the mid-latitude and subtropical glacier regions get a bit more weight because they have a larger sensitivity associated with a wetter climatic regime. This is more apparent when the cumulative glacier area and sensitivity are considered, starting at the North Pole (*figure 10.6*). For a good comparison, the total area and sensitivity are scaled to unity. Apparently, half of the area of glaciers and ice caps is found north of 70°N. With regard to the contribution to the global mean sensitivity the 50%-latitude is about 60°N. The difference is caused by the relatively large climate sensitivity of the glaciers in Alaska, Iceland and Norway. For the 75%-threshold the latitudes are 52°N for area and 36°N for climate sensitivity.

Figure 10.6 also makes it clear that, from a global perspective, glaciers between 30°N and 40°S are insignificant. The glaciers in Patagonia and on the south Atlantic islands contribute about 5% to the total area and about 10% to the global sensitivity.

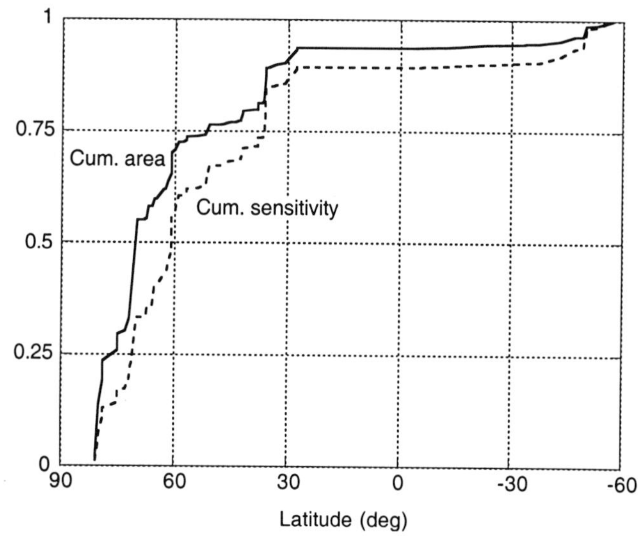

Figure 10.6. Cumulative area and climate sensitivity (to temperature) for glaciers and ice caps. The Greenland ice sheet is not included, but the surrounding glaciers and ice caps are.

10.3 Calculating the total contribution of glaciers to sea-level rise

The worldwide retreat of glaciers has frequently been connected with sea-level rise. However, it is not easy to quantify the total loss of ice. Pioneering work was done by Meier in 1984 [6]. He estimated that during the period 1900-1961 glaciers contributed 2.8 cm to sea-level rise. This estimate was based on an extrapolation of available mass balance data, including a scaling procedure in which glaciers with a larger mass turnover lost more ice (which is in agreement with the idea that glaciers in wet climates are more sensitive). More recent work along the same lines suggests a value of 0.7 cm for the period 1961-1990 [7].

Glaciers on which mass balance observations have been carried out are concentrated in a few regions and it is hard to see how an extrapolation to all glaciers and ice caps can be accurate. An alternative to extrapolating mass balance observations is to use the modelled sensitivity for the glacier regions described above in combination with meteorological records. This approach was followed by Zuo and Oerlemans [4] for the period 1865-1990, using temperature records. Here we will discuss some results briefly and make a comparison with the work of Dyurgerov and Meier [7] for the period 1961-1990.

The methodology is static in the sense that changes in the geometry of glaciers are not considered. A distinction is made between the effects of summer and 'non-summer' temperature anomalies (ΔT_s and ΔT_{ns}, respectively), because this makes it possible to capture the basic difference between maritime and dry glaciers (see *figure 4.13*). Ice volume for region k relative to an initial state in the year 1865 is then calculated from

$$V_k(t) = A_k \sum_{t'=1865}^{t} \left\{ (\Delta T_{s,k} - \theta)\, C_{Ts,k} + (\Delta T_{ns,k} - \theta)\, C_{Tns,k} \right\} . \tag{10.8}$$

The sensitivities of the mass balance for summer temperature and non-summer temperature are denoted by $C_{Ts,k}$ and $C_{Tns,k}$, respectively. Furthermore, a parameter θ is included as the initial imbalance between climatic conditions and state of the glaciers. In principle, θ is different for different places, but in this study only a global mean effect was allowed for.

For the calculation of temperature anomalies the initial reference period has to be specified. It is obvious that the result will depend on the choice of reference period. Zuo and Oerlemans (ZO) did calculations for reference periods 1865-1875, 1865-1895 and 1865-1915, but argue that the 1865-1895 period is the best choice.

As has been discussed earlier and was illustrated in *figure 9.1*, in almost all mountain regions in the world glaciers reached maximum stands in the 18th and 19th century. It appears that for the period 1865-1895 glaciers were slightly too large for the climatic conditions at the time (corresponding to $\theta < 0$). In the light of this observation three cases were considered: $\theta = 0$; -0.15; -0.3 K. For a discussion about the way in which $C_{Ts,k}$ and $C_{Tns,k}$ were obtained and in what way the temperature series were constructed from the World Climate Disc [8] the reader is referred to the original study [4]. Here we discuss a few results.

Applying eq. (10.8) to all glacier regions (but not the glaciers and small ice caps on Greenland, referred to as item 101 in *table 10.2*) and adding together the results yields the dashed curve in *figure 10.7*. For the calculation shown $\theta = 0$, and the reference period is 1865-1895. The contribution to sea-level change is shown together with the global mean temperature anomaly (T'). If the global mean temperature had been used as input, the curves would have been exactly the same. But they are not, so we conclude that the distinction between summer and non-summer temperatures as well as the regional differentiation in the temperature anomaly have a considerable impact on the final result. We will come back to this later.

The calculation suggests that the contribution of glaciers to sea-level rise was largest in the period 1925-1965 and after 1975, whereas values for the period 1965-1975 are rather low.

Is there a way of validating these results, for instance by looking directly at mass-balance observations? It is possible to compare the results to the global glacier melt calculated by Dyurgerov and Meier [DM, 7] for the period 1961-1990.

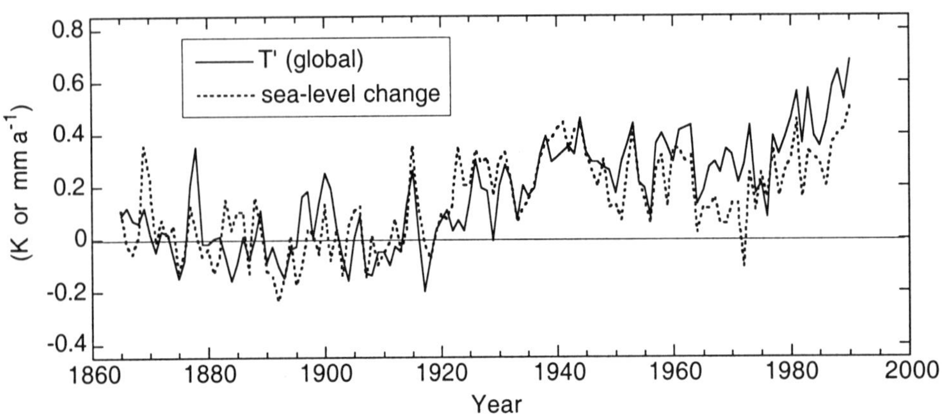

Figure 10.7. A comparison of the global mean temperature record and the calculated contribution of glaciers and ice caps to sea-level change for the period 1865-1990.

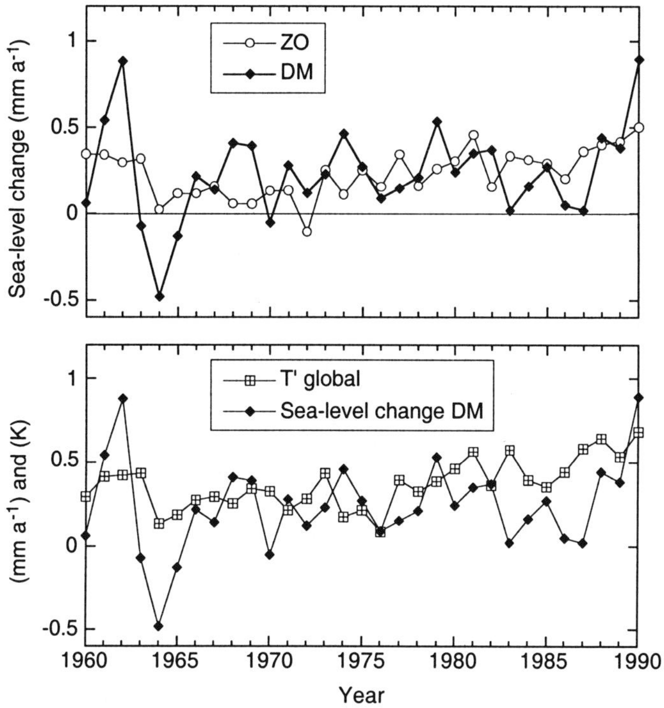

Figure 10.8. Upper panel: annual contribution of glaciers to sea-level change as calculated by Zuo and Oerlemans [ZO, 4] and Dyurgerov and Meier [DM, 7]. Lower panel: the DM curve compared with the global mean temperature anomaly (T').

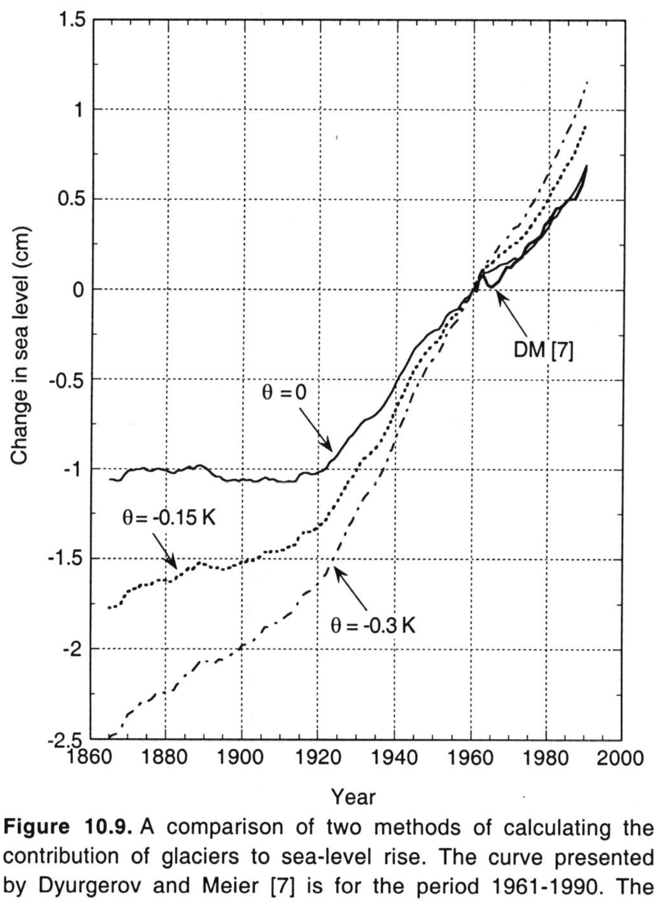

Figure 10.9. A comparison of two methods of calculating the contribution of glaciers to sea-level rise. The curve presented by Dyurgerov and Meier [7] is for the period 1961-1990. The other curves are from the study by Zuo and Oerlemans [4,9].

Figure 10.8 shows sea-level change due to glacier melt as calculated by DM and ZO. The correlation is very weak ($r = 0.37$) and hardly significant. The mean values for the period 1961-1990 are almost the same, however: 0.23 mm a^{-1} for ZO and 0.24 mm a^{-1} for DM. A comparison of the DM curve with the global mean temperature anomaly (*figure 10.8*) also reveals a weak correlation ($r = 0.38$). Although it is clear that global changes in precipitation may also explain part of the variance in the mass-balance record, such a weak correlation with temperature is surprising. As argued in section 9.7, there is not necessarily a good correlation on longer time scales, but for the year-to-year variability we expect to see a closer relation. This makes us wonder to what extent mass-balance observations on a limited number of glaciers can be used for an extrapolation to the global scale. Alternatively, we cannot rule out the possibility that the global mean temperature is a poor predictor of global mean mass balance.

From this discussion we can conclude that there is no way of judging which estimate (ZO or DM) is more reliable. Fortunately, a look at the cumulative contribution to sea-level rise from both methods yields rather similar results, as shown in *figure 10.9*. Three curves from the ZO calculation are shown, the only difference being the initial imbalance as prescribed by θ (see eq. 10.8). For negative values of θ, implying that in the late 19th century glaciers were too large for the prevailing climatic conditions, the calculated contribution to sea-level rise is of course larger. The curves have been drawn in such a way that they are all zero in the year 1960.

Although it is tempting to conclude that $\theta = 0$ is the best case because it leads to a good match between the ZO and DM calculations, we should be careful here. Before making firm statements we should wait until more accurate methods become available to deal with the initial state on the one hand, and for extrapolating scarce measurements to the entire globe on the other hand.

10.4 A discussion of the Greenland Ice Sheet

Most glaciologists will argue that the physical characteristics of the Greenland Ice Sheet have little to do with those of glaciers. Nevertheless, when looking at the (scarce) balance measurements performed on parts of the ice sheet (*figure 10.10*), it becomes clear that there are marked similarities between these measurements and balance profiles observed on glaciers. Meteorological observations [11] have shown that the processes governing the exchange of energy between atmosphere and ice surface are not very different from what is encountered on glaciers and ice caps in mid-latitude and subpolar regions of the Northern Hemisphere. Katabatic flows are persistent and deliver heat to the surface. As on glaciers, the seasonal march of the albedo is very large.

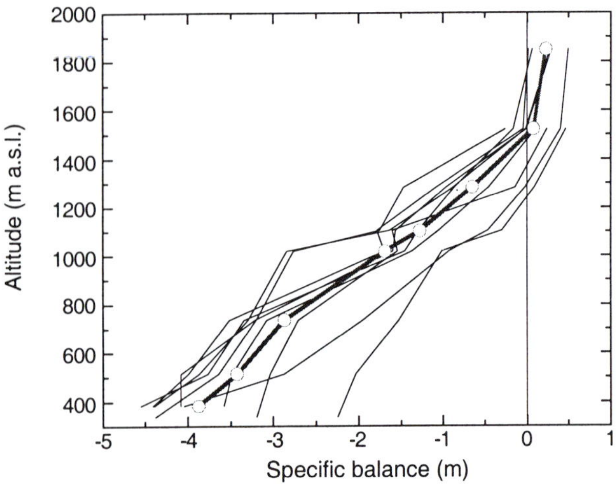

Figure 10.10. Balance profiles as measured by the Institute for Marine and Atmospheric Rsearch (Utrecht) in central West Greenland since 1990. Note that the measurements do not extend all the way to the ice divide. From [10] with additions. The red curve represents the mean profile over 9 years.

The abundance of melt water in some regions is probably related to the very small surface slope and the fact that the ice is not temperate [12]. The presence of melt water has implications because it lowers the albedo significantly [13].

In conclusion, the processes governing the shape of the balance profile are quite similar to those found on arctic ice caps and glaciers; this finding has stimulated the use of energy balance models to study the sensitivity of the Greenland Ice Sheet to climate change. In a first attempt the ice sheet was partitioned into four sectors and an energy/mass balance model was applied along four transects representing these sectors [14]. By using climatological data as input, the model predicted that about 60% of the total gain on the ice sheet is lost by melting and runoff, and, if the ice sheet is not too far out of balance (which is not certain at all), about 40% would be lost by calving.

A calculation of the spatial distribution of the mass-balance from the surface energy-balance, with temperature and precipitation fields as updated by Ohmura and Reeh [15] used as input, was first carried out by Van de Wal [16]. The calculations were done on a 20-km grid, which is just about good enough to resolve the ablation zone (but not in great detail). The result is shown in *figure 10.11*. An outstanding feature is the large zone of positive but small mass balance in the northeastern part of the ice sheet. This is the region with low precipitation and little melt. The southeastern part of the ice sheet is just above the main Atlantic storm track, which is reflected in the large amounts of precipitation, deposited when storms push up moist air masses from the east towards the ice divide. In this region the balance gradient is largest. The mass balance model predicts an ablation zone all round the ice sheet. This has now been confirmed by observations.

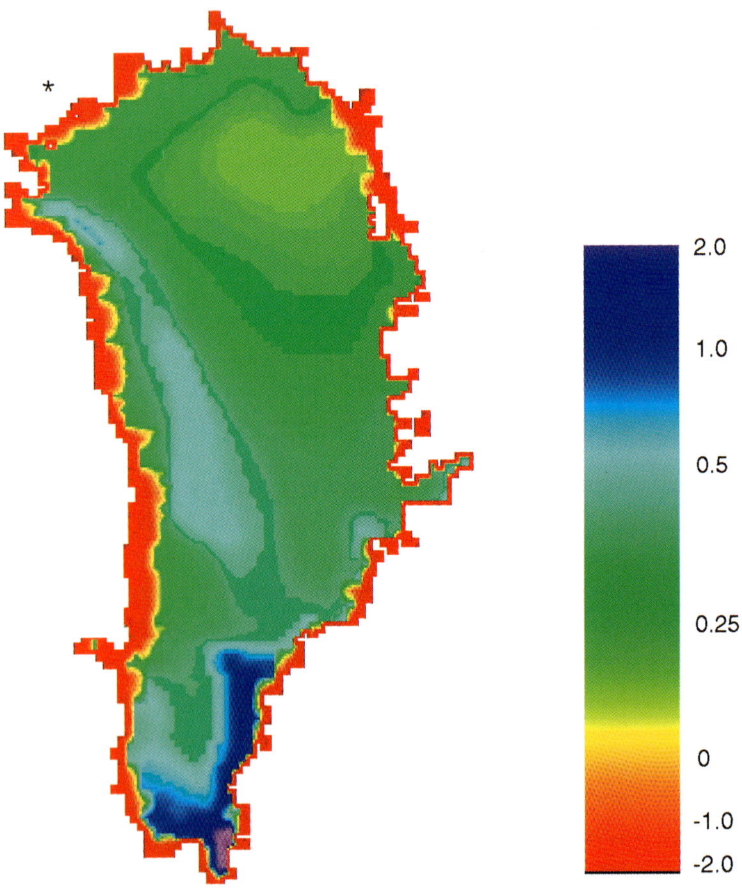

Figure 10.11. The specific balance rate (m a^{-1}) of the Greenland Ice Sheet as calculated with an energy-balance model from climatological input data [16]. Note that the scale is nonlinear. The grid resolution is 20 km.

A series of subsequent modelling studies has shed some light on the sensitivity of the mass balance of Greenland. Depending on the choice of the reference state and the details of the energy-balance model, values for C_T (sensitivity to annual mean temperature, section 4.7) range from -0.04 to -0.10 m a^{-1} K^{-1}. Recalling that the mean precipitation on the Greenland Ice Sheet is about 0.35 m a^{-1}, it appears that the sensitivity found still fits the scheme of section 4.7, where it was concluded that C_T is determined mainly by the precipitation regime (*figure 4.13*).

In view of the similarities between the mass-balance characteristics of Greenland and those of other glaciers and ice caps, ZO [4] have also applied to Greenland the static approach of calculating the contribution of ice melt to sea-level change. Four sectors were considered with different sensitivities for T_s and T_{ns}. But, as expected, for Greenland the sensitivity to temperature anomalies outside the summer is quite small. *Figure 10.12* shows that the contribution of the Greenland ice sheet to sea-level rise obtained in this way is similar to that of glaciers. Three curves are plotted to illustrate the effect of the initial imbalance. We conclude that the particular choice of θ has a significant effect on the result. Clearly, the contribution to sea-level rise was largest in the period 1920-1960.

To what extent does a regional differentiation lead to results that differ from a global approach in which a global mean temperature anomaly and a global mean glacier sensitivity are used? The result shown in *figure 10.7* already suggested that differences may be significant. Some additional calculations make it clear that this is indeed the case. In *figure 10.13* results are compared. The use of a global mean annual temperature anomaly gives a 25% smaller glacier melt. For Greenland the value is even halved. Apparently the temperature rise in regions with glaciers was larger than the global mean.

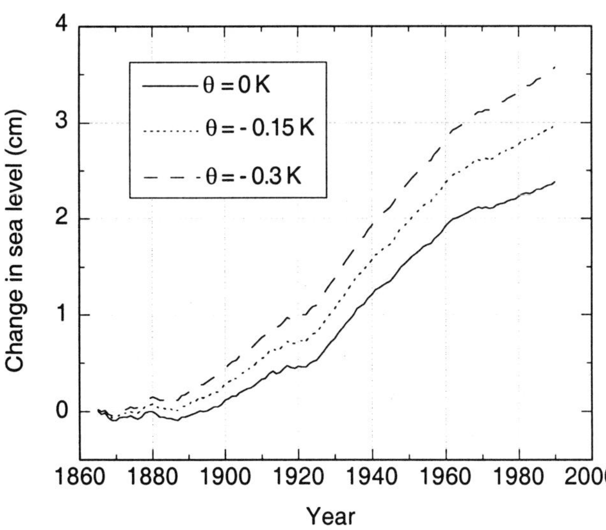

Figure 10.12. The contribution of the Greenland ice sheet to sea-level rise since 1865, calculated from temperature records [4]. The quantity θ is the initial imbalance (defined for the period 1865-1895).

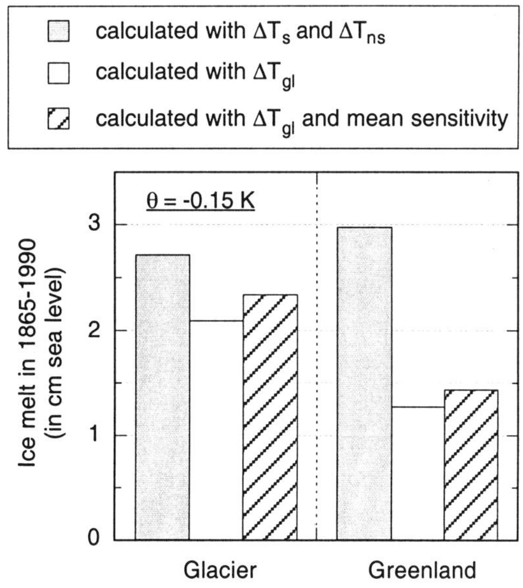

Figure 10.13. Different ways of calculating the contribution of glaciers and the Greenland Ice Sheet lead to different answers. This also applies to calculations with other values of θ (not shown) [4].

When a global mean sensitivity is used as well there are some additional changes, but these are smaller. Altogether, a regional differentiation is really to be preferred when one is studying the impact of glacier changes on global sea level.

10.5 A discussion of the Antarctic Ice Sheet

The Antarctic Ice Sheet differs from all other glaciers and ice sheets. Even in the warmest parts there is very little melt, let alone runoff. Modelling of the mass balance of the Antarctic ice sheet requires a different approach. It is the mesoscale and large-scale atmospheric dynamics that determines the amount of mass gained or lost every year rather than the local energy balance. The distribution of the specific balance is determined strongly by the precipitation pattern, although locally snow drift and sublimation play a significant role.

Attempts have been made to express the specific balance in terms of temperature, altitude, slope, distance from the coast, etc. Regression analysis has been applied to (scarce and inhomogeneous) data on accumulation to find the optimum schemes [17]. Although a great deal of the variance can be explained with such statistical methods, it is not clear that they are suitable for determining the sensitivity of the mass balance to climate change. One of the problems is the interdependence of predictors. Altitude, temperature and distance from the coast are certainly correlated. It is not surprising that in all analyses air temperature shows up as an important predictor. The idea that the total precipitation on the Antarctic Ice Sheet is determined to a considerable extent by the moisture-holding capacity of the overlying atmosphere was put forward a long time ago, and has received support from more detailed atmospheric modelling studies [18] and ice-core analysis [19].

In recent years general circulation models of the atmosphere have been used more and more to simulate the mass balance of the large ice sheets. It is mainly the increase in spatial resolution that has led to a considerable improvement of the simulated mass-balance fields. Hitherto, the most detailed modelling study with an atmospheric model is the one by Van Lipzig [20]. She carried out a 10-year hindcast (1980-1989) with a carefully calibrated, high-resolution regional climate model, driven by observed meteorological fields at the boundary of the model domain (at ~55°S) and observed sea-ice distributions. The boundary conditions were refreshed every six hours and taken from re-analyses from the European Centre for Medium Range Weather Forecasting (ECMWF, Reading, UK). With a spatial resolution of 55 km and a rather detailed treatment of the boundary layer, this model is able to reproduce the mass balance and meteorological fields over the Antarctic Ice Sheet in considerable detail.

In *figure 10.14* the 10-year specific balance from this hindcast is compared to the mass balance map of Bentley and Giovinetto [21]. The agreement is quite good, except for a few regions of negative balance. These are the result of strong sublimation simulated in mountain regions because of a large prescribed value for the surface roughness.

We should realise that the mass-balance map is based on unevenly distributed measurements done with very different methods and averaged over different periods of time. So perhaps we cannot expect a closer resemblance. In any case, it is clear that the specific balance field on the Antarctic Ice Sheet is of a totally different nature when compared to Greenland. On the Greenland Ice Sheet the balance increases when going inland, on the Antarctic ice sheet it decreases. So it does not come as a surprise that the sensitivity of the mass balance to temperature change is different for the two ice sheets. In *table 10.3* a few estimates are summarised.

If accumulation on the Antarctic ice sheet is assumed to be proportional to the saturation vapour pressure of the air, and if changes in sublimation are ignored, a typical sensitivity turns out to be 14 mm a^{-1} K^{-1}. This is considerably larger than the values found by regression analysis performed on the mass balance measurements.

With a 2-dimensional axisymmetric model, in which the katabatic outflow and large-scale inflow aloft are modelled explicitly, a value of 13 mm a^{-1} K^{-1} is obtained.

The 3-dimensional climate model discussed above yields a much larger value for C_T (24 mm a^{-1} K^{-1}) when the hindcast is run again with a temperature increase at the boundaries. The reason for this is the water vapour feedback acting in the atmosphere overlying the ice sheet. The temperature increase over the ice sheet turns out to be almost twice as large as the temperature increase at the boundary of the model domain in the southern ocean. If the temperature increase over the ice sheet is used to calculate C_T, a value of 14 mm a^{-1} K^{-1} is found.

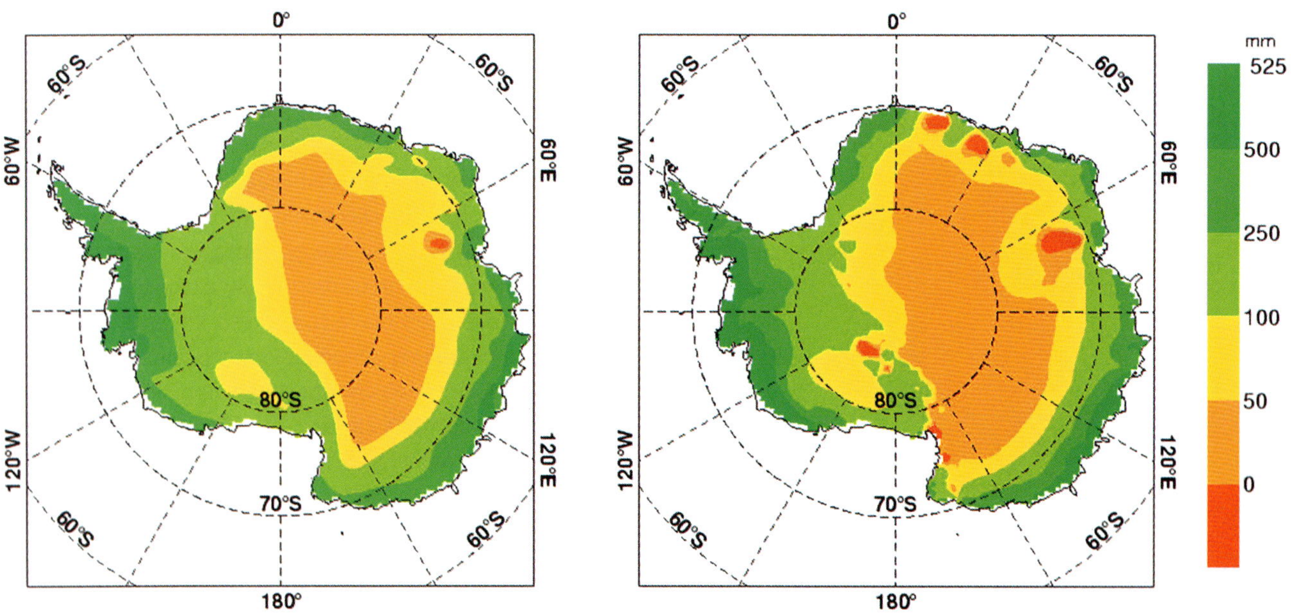

Figure 10.14. The specific balance of the Antarctic Ice Sheet (mm a-1) as observed [21] (left) and modelled with a high-resolution regional climate model [20] (right). The computed balance field originates from a 10-year hindcast for the period 1980-1989, driven by ECMWF re-analyses at the model boundaries. The resolution of the climate model is 55 km.

study		C_T (mm a^{-1} K^{-1})	sea level (mm a^{-1} K^{-1})
Antarctic Ice Sheet			
Fortuin and Oerlemans (1990) [17]	proportional to saturation vapour pressure	14	-0.71
Muszynski and Birchfield (1985) [17]	statistical analysis	7.5	-0.38
Giovinetto et al. (1990) [22]	statistical analysis	9	-0.46
Fortuin and Oerlemans (1990) [17]	statistical analysis	6	-0.20
Fortuin and Oerlemans (1992) [23]	2-D atmospheric model (axi-symmetric)	13	-0.27
Van Lipzig (1998) [20]	3-D atmospheric model (55 km resolution)	24	-0.50
Van Lipzig (1998) [20]	3-D atmospheric model (with T over ice sheet)	14	-0.29
Greenland Ice Sheet			
Oerlemans (1991) [14]	4-sector energy-balance model	-172	0.37
Huybrechts et al. (1991) [22]	degree-day model (20 km resolution)	-139	0.30
Van de Wal and Oerlemans (1994) [16]	energy-balance model (20 km resolution)	-139	0.30
Glaciers and ice caps			
see section 10.2	area-weighted mean sensitivity of 101 regions	-220	0.37

Table 10.14. The sensitivity of the mean specific balance of the Antarctic Ice Sheet to a uniform temperature change, as obtained in different studies. Values for the Greenland Ice Sheet and for glaciers and ice caps as discussed in section 10.2 are included for comparison. The last column shows the implication for sea-level change (assuming a constant surface area of the glaciers and ice sheets and no changes in iceberg calving or mass flux across grounding lines).

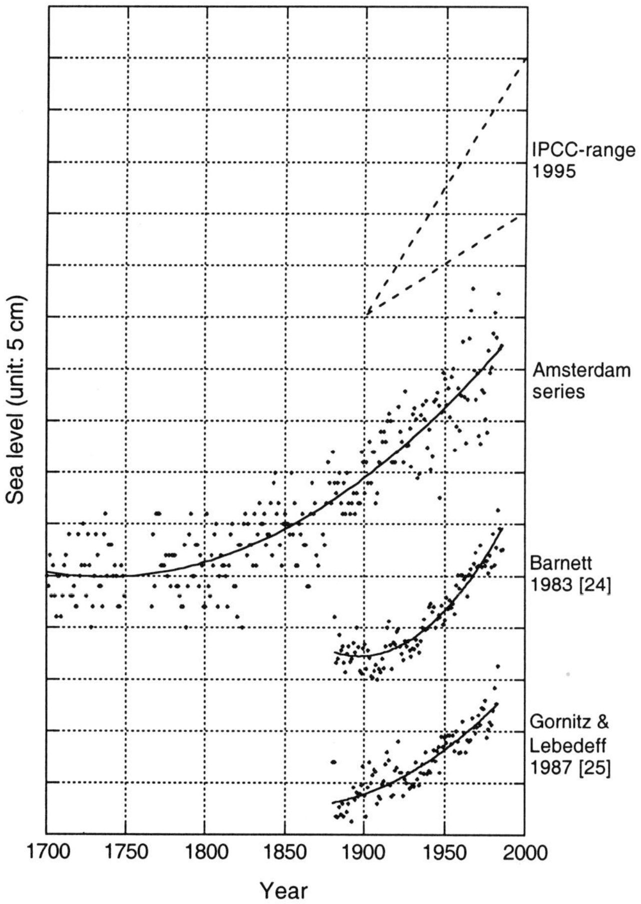

Figure 10.15. Global sea-level change. The longest record (Amsterdam) is shown together with two estimates of global mean sea-level change. The range suggested by IPCC [3] is shown by dashed lines.

10.6 Sea-level rise, land ice and thermal expansion

As discussed in section 10.2 sea level is currently rising at a rate of 10 to 25 cm per 100 years. Some data on sea-level change are summarised in *figure 10.15*. The longest record of sea-level is the Amsterdam record, which started around 1700. Amsterdam is certainly not on the most stable ground one can think of, and the record has been criticised for this reason. Nevertheless, some careful corrections have been carried out. In 1932 the record became meaningless because the open connection to the sea was disrupted. However, sea level was measured at many other Dutch coastal stations and the record has been extended with data from Den Helder. The period of overlap is about 100 years and the correlation between the two records high, so the records can be easily combined.

The records of Barnett [24] and Gornitz and Lebedeff [25] give annual values for the global mean sea level. These records are based essentially on the same data set, but use different methods to correct for local effects of tectonic or isostatic origin. A second-order polynomial is shown for all data. There is some indication in these records (and others) that sea-level rise has been accelerating. However, this idea is by no means generally accepted.

Many more studies have been done with sea level data, some including sophisticated methods (geodynamic models) to correct for local movements. IPCC has summarised such studies and came to the conclusion that global mean sea-level rise over the last 100 years has been between 10 and 25 cm, with a best estimate of 18 cm.

Explaining this rise is one of the key issues in climate change research. Projections of sea-level rise for the coming centuries have been presented on many occasions, but it is a fact that the current change is not well understood. Among the most important factors mentioned are: thermal expansion of the oceans, changes in the volume of land ice, changes in ground water storage, lakes and hydropower reservoirs, etc. Very little is known about possible changes in deep groundwater reservoirs. Thermal expansion and land ice changes are generally considered to be more important than the other factors. However, the possibility that part of the signal comes from long-term geodynamic processes cannot be ruled out.

With regard to thermal expansion, to date there has been only one study in which an ocean model is forced by observed sea surface temperatures [26]. It appears that thermal expansion has made a significant contribution to sea-level rise. The precise answer depends on model parameters like overturning speed and diffusivity in the ocean, but the order-of magnitude is the same as the contribution of glaciers and Greenland. In *figure 10.16* the various contributions are summarised. For glaciers and Greenland the results for $\theta = -0.15$ K have been used. We note that the total rise in sea level, explained in this way, is about 10 cm since the end of the 19th century. This is much less that observed, or more precisely, at the very lower limit of the range given by IPCC.

A weak point in all of this is the choice of the initial state. The ocean was assumed to be in balance in the year 1890, and for the glacier and Greenland a small imbalance was used. But what about long-term changes in the ice sheets of Greenland and Antarctica?

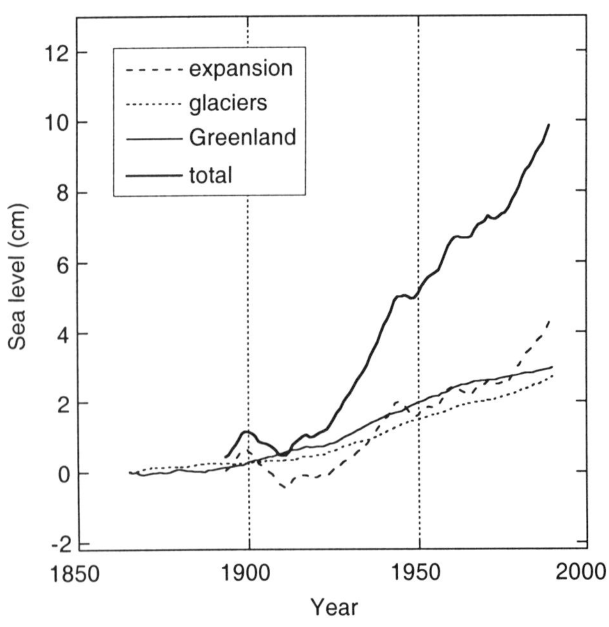

Figure 10.16. An estimate of the contributions of thermal expansion and glacier melt to sea-level rise [4, 26].

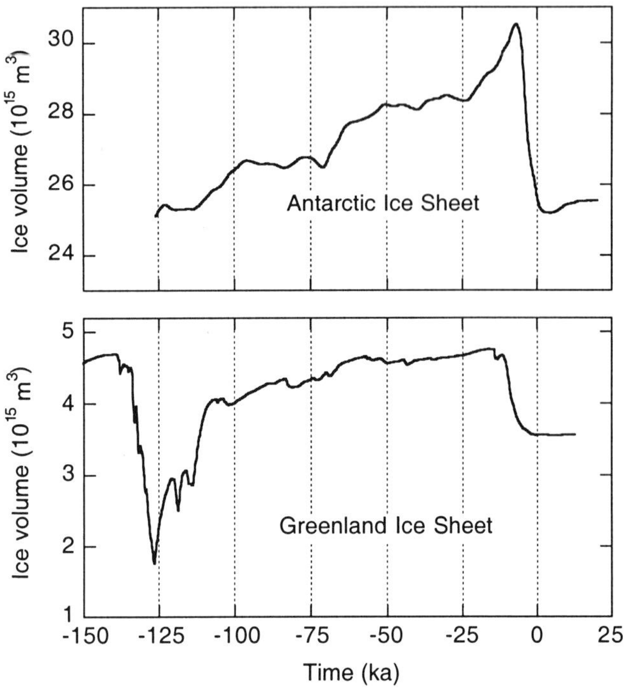

Figure 10.17. Simulated ice volume of the Antarctic and Greenland ice sheets through a full glacial cycle [27, 28]. The present time is zero.

In fact, current data sets do not let us determine with any reliability the net balance of the ice sheets (and not even the sign). One possible way out is to try to simulate the temporal evolution of the ice sheets with comprehensive models and look at the outcome for the present state of balance. This only makes sense if the time span considered is at least a few times the response time of the ice sheets, and if appropriate forcing functions can be defined. In this approach, model parameters can be varied to match constraints from observational studies in the best possible way. Such constraints are for instance approximate grounded ice positions derived from marine glacial deposits, data on local vertical land movements (probably in combination with information about relative sea-level stands), and, last but not least, the rapidly increasing information from ice cores and boreholes.

Figure 10.17 shows the result of two studies with fairly comprehensive ice-sheet models [27, 28]. In these models ice mechanics and thermodynamics are fully coupled, and information from ice cores (paleotemperature) and reef terraces (sea level) were used as forcing functions. The earlier parts of the simulated curves depend largely on how the initial state is chosen. However, for the last 50 ka or so this influence should have been damped out. The model simulations for the Antarctic and Greenland ice sheets both imply that a great deal of ice was lost during the last deglaciation, although the loss may be less than suggested in earlier studies (in particular for the Antarctic ice sheet [29]). The Antarctic ice sheet responds mainly to sea-level change.

Significant grounding-line retreat is initiated by the sea-level rise due to the melting of the Northern Hemisphere ice sheets. Change in accumulation, coupled to changes in temperature, is only of secondary importance. In contrast, for the Greenland Ice Sheet increasing temperatures during the deglaciation lead to a strongly negative balance in the ablation zones, leading to the retreat and decrease in volume.

The simulations suggest that at present the Greenland Ice Sheet is close to a state of balance, whereas the Antarctic Ice Sheet is still losing mass. In terms of sea-level change, the contributions are +1.15 mm a^{-1} and -0.02 mm a^{-1}, respectively. It is clear that such values are subject to a large uncertainty. If, for instance, the phasing of Antarctic deglaciation in the model is slightly late, the modelled contribution to sea-level rise may be halved or even close to zero. Nevertheless, in principle dynamic modelling of the ice sheets is an attractive approach, because in the near future models will become more accurate and constraints more definite.

References and further reading

1. O. Sigurdsson (1998): Glacier variations in Iceland 1930-1995. From the database of the Icelandic Glaciological Society. *Jökull* **45**, 3-26.

2. J. Oerlemans (1994): Quantifying global warming from the retreat of glaciers. *Science* **264**, 243-245.

3. W. Haeberli, H. Bösch, K. Scherler, G. Østrem and C.C. Wallén (1989): *World Glacier Inventory. Status 1988.* IAHS(ICSI)-UNEP-UNESCO.

   IPCC (1996): *Climate Change 1995. The Science of Climate Change* (WMO-UNEP). Cambridge University Press, 572 pp.

4. Z. Zuo and J. Oerlemans (1997): Contribution of glacier melt to sea-level rise since AD 1865: a regionally differentiated calculation. *Climate Dynamics* **13**, 835-845.

5. J. Oerlemans and J.P.F. Fortuin (1992): Sensitivity of glaciers and small ice caps to greenhouse warming. *Science* **258**, 115-117.

6. M.F. Meier (1984): Contribution of small glaciers to global sea level. *Science* **226**, 1418-1421.

7. M.B. Dyurgerov and M.F. Meier (1997) Year-to-year fluctuations of global mass balance of small glaciers and their contribution to sea-level changes. *Arctic and Alpine Research* **29**(4), 392-401.

8. World Climate Disc (Chadwyck-Healey Ltd, 1992).

9. J. Oerlemans (1999): Comments on "Mass balance of glaciers other than the ice sheets" by Cogley and Adams. *Journal of Glaciology* **45** (150), 397-398.

10. R.S.W. van de Wal, R. Bintanja, W. Boot, M.R. van den Broeke, L.A. Conrads, P.G. Duynkerke, P. Fortuin, E.A.C Henneken, W. Knap, M. Portanger, H.F. Vugts and J. Oerlemans (1995): Mass balance measurements in the Søndre Strømfjord area in the period 1990-1994. *Zeitschrift für Gletscherkunde und Glazialgeologie* **31**, 57-63.

11. J. Oerlemans and H.F. Vugts (1993): A meteorological experiment in the melting zone of the Greenland ice sheet. *Bulletin of the American Meteorological Society,* **74**, 355-365.

    M.R. van den Broeke, P.G. Duynkerke and E.A.C. Henneken (1994): Heat, momentum and moisture budgets of the katabatic layer over the melting zone of the West Greenland ice sheet in summer. *Boundary-Layer Meteorology* **71**, 393-413.

    R.S.W. van de Wal and A.J. Russell (1994): A comparison of energy balance calculations, measured ablation and meltwater runoff near Søndre Strømfjord, West Greenland. *Global and Planetary Change* **9**, 29-38.

    P.G. Duynkerke and M.R. van den Broeke (1994). Surface energy balance and katabatic flow over glacier and tundra during GIMEX-91. *Global and Planetary Change* **9**, 17-28.

    W. Greuell and T. Konzelmann (1994): Numerical modelling of the energy balance and the englacial temperature of the Greenland ice sheet: calculations for the ETH-Camp location (West-Greenland, 1155 m a.s.l.). *Global and Planetary Change* **9**, 91-114.

    T. Konzelmann, R.S.W. van de Wal, W. Greuell, R. Bintanja, E.A.C. Henneken and A. Abe-Ouchi (1994): Parameterization of global and longwave incoming radiation for the Greenland ice sheet. *Global and Planetary Change* **9**, 69-78.

    M.R. van den Broeke (1996): Characteristics of the lower ablation zone of the west Greenland ice sheet for energy-balance modelling. *Annals of Glaciology* **23**, 160-166.

12. Z. Zuo and J. Oerlemans (1996): Modelling albedo and specific balance of the Greenland ice sheet: calculations for the Søndre Strømfjord transect. *Journal of Glaciology* **42**, 305-317.

13. W.H. Knap and J. Oerlemans (1996): The surface albedo of the Greenland ice sheet: satellite-derived and in situ measurements in the Søndre Strømfjord area during the 1991 melt season. *Journal of Glaciology* **42**, 364-374.

14. J. Oerlemans (1991): The mass balance of the Greenland ice sheet: sensitivity to climate change as revealed by energy-balance modelling. *The Holocene* **1**, 40-49.

15. A. Ohmura (1987): New temperature distribution maps for Greenland. Zeitschrift für Gletscherkunde und Glazialgeologie **23** (1), 1-45.

A. Ohmura and N. Reeh (1991): New precipitation and accumulation maps for Greenland. Journal of Glaciology **37** (125): 140-148.

16. R.S.W. van de Wal (1992): *Ice and Climate.* Ph. D. Thesis, Utrecht University.

R.S.W. van de Wal and J. Oerlemans (1994): An energy balance model for the Greenland ice sheet. *Global and Planetary Change* **9**, 115-131.

17. J. Oerlemans (1982): A model of the Antarctic Ice Sheet. *Nature* **297**, 550-553.

I. Muszynski and G.E. Birchfield (1985): The dependence of Antarctic accumulation rates on surface temperature and elevation. *Tellus* **37A**, 204-208.

M.B. Giovinetto, N.M. Waters and C.R. Bentley (1990): Dependence of Antarctic surface mass balance on temperature, elevation and distance to open water. *Journal of Geophysical Research* **95**, 3517-3531.

J.P.F. Fortuin and J. Oerlemans (1990): Parameterization of the annual surface temperature and mass balance of Antarctica. *Annals of Glaciology* **14**, 78-84.

18. C. Genthon (1994): Antarctic climate modelling with general circulation models of the atmosphere. *Journal of Geophysical Research* **99** (D6), 12953-12961.

19. C. Lorius, J. Jouzel, D. Raynaud, J. Hansen and H. Le Treut (1990): The ice-core record: climate sensitivity and future greenhouse warming. *Nature* **347**, 139-145.

20. N.P.M. van Lipzig (1999): *The surface mass balance of the Antarctic Ice Sheet: a study with a regional atmospheric model.* Ph.D. Thesis. Utrecht University.

21. M.B. Giovinetto and C.R. Bentley (1985): Surface balance in ice drainage systems of Antarctica. *Antarctic Journal of the United States* **4**, 6-13.

22. Ph. Huybrechts, A. Letreguilly and N. Reeh (1991): The Greenland ice sheet and greenhouse warming. *Paleogeography, Paleoclimatology, Paleoecology* **89**, 79-92.

23. J.P.F. Fortuin and J Oerlemans (1992): An atmospheric model for simulating the mass balance and temperature on the Antarctic ice sheet. *Zeitschrift für Gletscherkunde und Glazialgeologie* **26** (1), 31-56.

24. T.P. Barnett (1983): Recent changes in sea level and their possible causes. *Climatic Change* **5**, 15-38.

T.P. Barnett (1984): The estimation of "global" sea level change: a problem of uniqueness. Journal of Geophysical Research **89** (C5), 7980-7988.

25. V. Gornitz and S. Lebedeff (1987): Global sea level changes during the past century. In: *Sea level change and coastal evolution* (eds.: D. Nummedal, O.H. Pilkey and J.D. Howard), SEPM Publ. no. 41, 3-16.

26. J.R. De Wolde, R. Bintanja and J. Oerlemans (1995): On thermal expansion over the last hundred years. *Journal of Climate* **8** (11), 2881-2891.

27. Ph. Huybrechts (1990): A 3-D model for the Antarctic ice sheet: a sensitivity study on the glacial-interglacial contrast. *Climate Dynamics* **5**, 79-92.

28. Ph. Huybrechts (1994): The present evolution of the Greenland ice sheet: an assessment by modelling. *Global and Planetary Change* **9**, 39-51.

R.S.W. van de Wal (1999): Processes of buildup and retreat of the Greenland ice sheet. *Journal of Geophysical Research* **104** (D4), 3899-3906.

11. The future of glaciers

11.1 Introduction

The large sensitivity of glaciers to climate change implies that the projected greenhouse warming may have far-reaching consequences for their distribution. Glaciers that span a small altitudinal range will disappear more quickly than glaciers that have accumulation zones far above the present day equilibrium line. Flat ice caps, such as those found in Scandinavia, are the most vulnerable.
In a warmer world the number of glaciers and ice caps will not necessarily be smaller than today. Actually, in the marginal zones of the large arctic ice caps and the Greenland Ice Sheet more valley glaciers may form as a result of general ice wastage! But, although the Antarctic Ice Sheet may in fact grow, the volume of glaciers and ice caps will most likely decrease.
With regard to the issue of future sea-level rise, researchers have sought methods to extrapolate modelling results for individual glaciers to all glaciers and ice caps in the world. The simplest way is to lump all glaciers together into a single equation describing the loss of ice. The author may have been the first to try this approach more than ten years ago; his equation was based on the wish to do something very simple rather than on actual evidence [1]. The equation reads:

$$\frac{dV}{dt} = -c_1 \, (T - T_{ref}) \left\{ V - V_{ref} \exp\left[\frac{-(T - T_{ref})}{c_2} \right] \right\} . \qquad (11.1)$$

Here V is ice volume and T global mean temperature at time t, V_{ref} is the ice volume in the reference state which corresponds to the temperature T_{ref}, and c_1 and c_2 are constants.
First of all we note that according to this formulation the equilibrium volume of all glaciers and ice caps drops off exponentially with the global mean temperature. This may be a reasonable assumption, but it is not easy to determine a value for c_2 (in the earlier study the author used $c_2 = 4.5$ K [1]). The time scale for adjusting the ice volume to the prevailing climatic conditions is $[c_1(T - T_{ref})]^{-1}$, which implies that the response of glaciers is faster and more pronounced when the temperature perturbation $(T - T_{ref})$ is larger [in a purely linear model the time scale would be a constant].
Meanwhile a number of glaciers have been modelled explicitly. A range of climate-change scenarios has been imposed on these models in order to find out about the vulnerability of glaciers to enhanced greenhouse warming. It is clear that such studies are needed to construct a model that describes the global picture in an adequate way. Possibly the most systematic study to date was the one carried out in the framework of EISMINT (European Ice-Sheet Modelling Initiative, a programme launched and coordinated by the European Science Foundation). We will discuss this study [2], because it has given us considerable insight into the possible response of glaciers to climate warming. We will look at the role of a glacier's geometry, which leads to large differences in the response of various glaciers. Then we will also return to the question of whether all glaciers in the world can be lumped into a single equation like eq. (11.1).

11.2 The EISMINT-study of glaciers in the 21st century

The EISMINT-study was set up in a simple way. Researchers using numerical models for particular glaciers were asked to run their models until the year 2100 with a set of prescribed climate-change scenarios. Dynamic calibration with historical records of glacier length, as described in section 8.5, was applied whenever this was possible. In the end 11 researchers took part in the study, in which 12 glaciers were investigated. The results were compared at a workshop in October 1996, which took place at the Landgasthof Morteratsch, Pontresina, Switzerland. A summary of the results was published [2]; a selection of the material is discussed below.

glacier / ice cap	area (km^2)	ice volume (km^3)	Ela (m a.s.l.)	first info (year)
Franz Josef Glacier (New Zealand) #	34	4.89	1650	1750
Glacier d'Argentière (France) # ($)	15.6	1.91	2900	1590
Haut Glacier d'Arolla (Switzerland) # $	6.3	0.33	3200	1933
Hintereisferner (Austria) # $	7.4	0.44	2950	1770
Nigardsbreen (Norway) # $	48	3.8	1550	1700
Pasterze (Austria) # ($)	19.8	2.62	2880	1605
Rhonegletscher (Switzerland) # $	17.7	2.58	2930	1602
Storglaciären (Sweden) # $	3.1	0.30	1460	1897
Unt. Grindelwaldgl. (Switzerland) #	21.7	1.83	2770	1534
Blondujökull (Iceland) ($)	226	46.9	1300	recent
Illvidrajökull (Iceland) ($)	116	25.8	1250	recent
KGI ice cap (Antarctica) ($)	1402	155	100	recent

Table 11.1. Glaciers and ice caps considered in the EISMINT-study [2]. Blöndujökull and Illviðrajökull are part of Hofsjökull (Iceland). KGI ice cap refers to the King George Island ice cap, which is used here as a name for the entire land-ice cover on King George Island, Antarctica Peninsula. The third column gives the (estimated) equilibrium-line altitude (Ela). The ice volumes given are modelled volumes and may deviate somewhat from the real volumes.
Further information: # historical length record available (see last column for first data point), $ mass balance observations available, ($) limited amount of mass balance measurements available (a few years and/or covering a small part of the glacier only).

The glaciers that were studied are listed in *table 11.1*. With regard to glacier volume and glacier area, three orders of magnitude are covered. The amount and type of data available vary considerably from glacier to glacier and therefore some models could be calibrated more thoroughly than others.

Most of the glaciers listed are valley glaciers. However, Blöndujökull and Illviðrajökull are sectors of an ice cap, whereas KGI ice cap applies to the entire ice cover on King George Island (Antarctic Peninsula). All the glaciers and ice caps are assumed to be essentially temperate with calving occurring only at KGI. In the model for KGI ice cap, iceberg calving was treated in a very simple way: at the coastline ice thickness was simply set to zero [3].

Although the models are not identical, they are based on the assumption that ice velocity is determined by the local driving stress, as described in Chapter 5. For KGI ice cap the calculations were done on a two-dimensional grid, for the other glaciers they were done along flowlines. The mass balance was treated in different ways. For glaciers for which good mass balance observations and climate data exist, calibrated mass balance models (based on a degree-day or an energy balance approach) were employed. In other cases a more schematic formulation had to be used.

For most glaciers it was possible to calibrate the flow models with historical records of glacier length. Some of these records go far back in time, as indicated in the last column of *table 11.1*. An interesting result of this procedure is the projected glacier length in the case of no climatic change. One can assume that mass-balance conditions remain as they were for the period 1961-1990 (in this study 1990 was taken as the reference year for all calculations), and consider the future behaviour of the glaciers. This sheds light on the current state of balance of these glaciers (*figure 11.1*). It should be noted that this 'climatological mass balance' is determined largely by the calibration (except for Nigardsbreen and Storglaciären; in the period 1961-1990, mass-balance observations are available for 28 years for Nigardsbreen and for 30 years for Storglaciären).

If we look at the general picture emerging from *figure 11.1* it appears that in the coming decades most glaciers in the sample will continue to retreat. During the period 1961-1990 these glaciers were therefore larger than the equilibrium sizes that correspond to the prevailing climatic conditions. The most notable retreats are predicted for the Pasterze, Hintereisferner and Haut Glacier d'Arolla. These glaciers are in really bad shape and will most certainly lose a considerable amount of mass in the coming decades. For three glaciers, (Franz-Josef Glacier, the Untere Grindelwaldgletscher and Nigardsbreen) significant advances are predicted. The reason

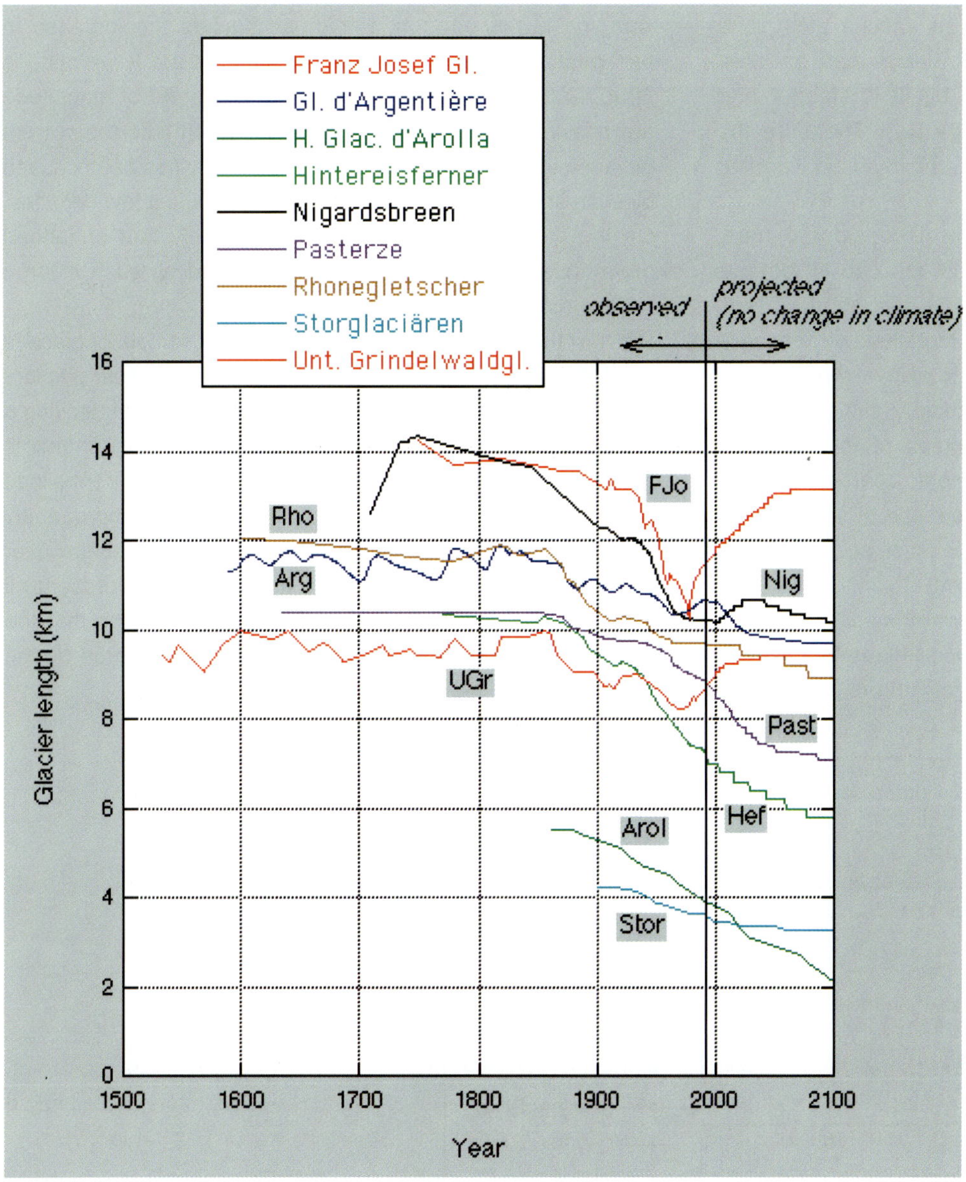

Figure 11.1. Records of glacier length and projection into the future, obtained by calibration and running models with climatic conditions set equal to those over the period 1961-1990 [2].

for this is the positive mass-balance perturbation over the period 1961-1990. For Franz-Josef Glacier as well as the Untere Grindelwaldgletscher the positive mass balance has resulted from the dynamic calibration (otherwise the models could not have reproduced the advance that occurred in the nineteen-eighties). For Nigardsbreen the positive mass balance is shown in the measurements (it is caused mainly by large quantities of precipitation).

Altogether, the results shown in *figure 11.1* make it clear that around 1990 most of the glaciers studied were not in equilibrium with the prevailing climatic conditions. Although the collection of glaciers used is rather small, this statement probably applies to most larger glaciers in the world.

In the EISMINT-study the period 1961-1990 was chosen as a refence period. Although this officially constitutes a period for which climate statistics can be evaluated, it should be realised that some ambiguity is involved. For some glaciers the period 1961-1990 may have been a relatively warm period, for other glaciers it may have been a cold period.

Next we turn to the climatic change experiments. In the EISMINT-study, six climate-change scenarios were used. These scenarios were fairly simple: constant warming rates of 0.01, 0.02 and 0.04 K a^{-1} were imposed for the period 1990 - 2100. All the calculations were repeated with an additional 10% change in precipitation per degree warming. We refer to the six scenarios as 0.01, 0.01^+, 0.02, 0.02^+, 0.04 and 0.04^+. In *Figure 11.2* results are plotted for the 0.02-scenario. Shown is the ice volume, normalised with the 1990-value. As expected, the differences are large. Haut Glacier d'Arolla, a small glacier already in a strongly retreating mode, will have virtually disappeared by the year 2060. In general the smaller glaciers lose relatively more mass.

By the year 2100, the larger glaciers and ice caps will have lost typically 40% of their volume. Nigardsbreen is an exception. In spite of its size and the fact that the glacier is currently advancing, it loses a large amount of mass. This is related to the hypsometry of Nigardsbreen: the bulk of the glacier is resting on a plateau (*figure 7.1*). Consequently, modest warming turns a large part of the accumulation zone into an ablation zone. Franz-Josef Glacier has an area comparable to that of Nigardsbreen but it is steeper and covers a wider altitudinal range (*figure 8.1*). Therefore the response of the Franz-Josef Glacier to the imposed warming is less pronounced.

It is striking that the ice cap (sectors) considered here, Blöndujökull, Illviðrajökull, and the KGI ice cap all behave similarly: by 2100 they are likely to have lost about 35% of their volume. It should be noted that integrations for these ice caps start from an equilibrium state. Nevertheless, the fractional loss of ice follows a remarkably similar pattern. Altogether, it is not possible to find a simple relation between the fractional loss of ice and the geometric characteristics of the glaciers for which calculations were made. It appears that the hypsometry of a particular glacier must be considered explicitly if ice wastage is to be estimated.

The melting of ice due to climatic warming is not easily compensated for by an increase in precipitation. As shown in *figure 11.3*, a 10% increase in precipitation per degree warming is not enough. For most glaciers the increase in precipitation needed to cancel the effect of a 1 K-warming is 30 to 40%. Precipitation increases of this order of magnitude are possible on a local or regional scale. In fact, the current advance of glaciers in Southwestern Norway has been caused by such an increase in precipitation over the last few decades. On the global scale, however, it is unlikely that enhanced greenhouse warming will be accompanied by such a large increase in mass accumulation on glaciers.

We now need to find out to what extent the sample of glaciers used in the EISMINT-study is representative for all glaciers and ice caps in the world. There are several ways of deriving a normalised global response curve from the results shown in *figure 11.2*. One approach is simply to average all normalised individual response curves. We refer to the resulting quantity as $<V_{sc}>$ (sc = scaled with 1990-volume; < > is mean over sample). If the resulting curve is considered to represent all glaciers in the world, the implicit assumption will be that the total glacier areas in each 'size class' as represented by the glaciers in *table 11.1* are equal.

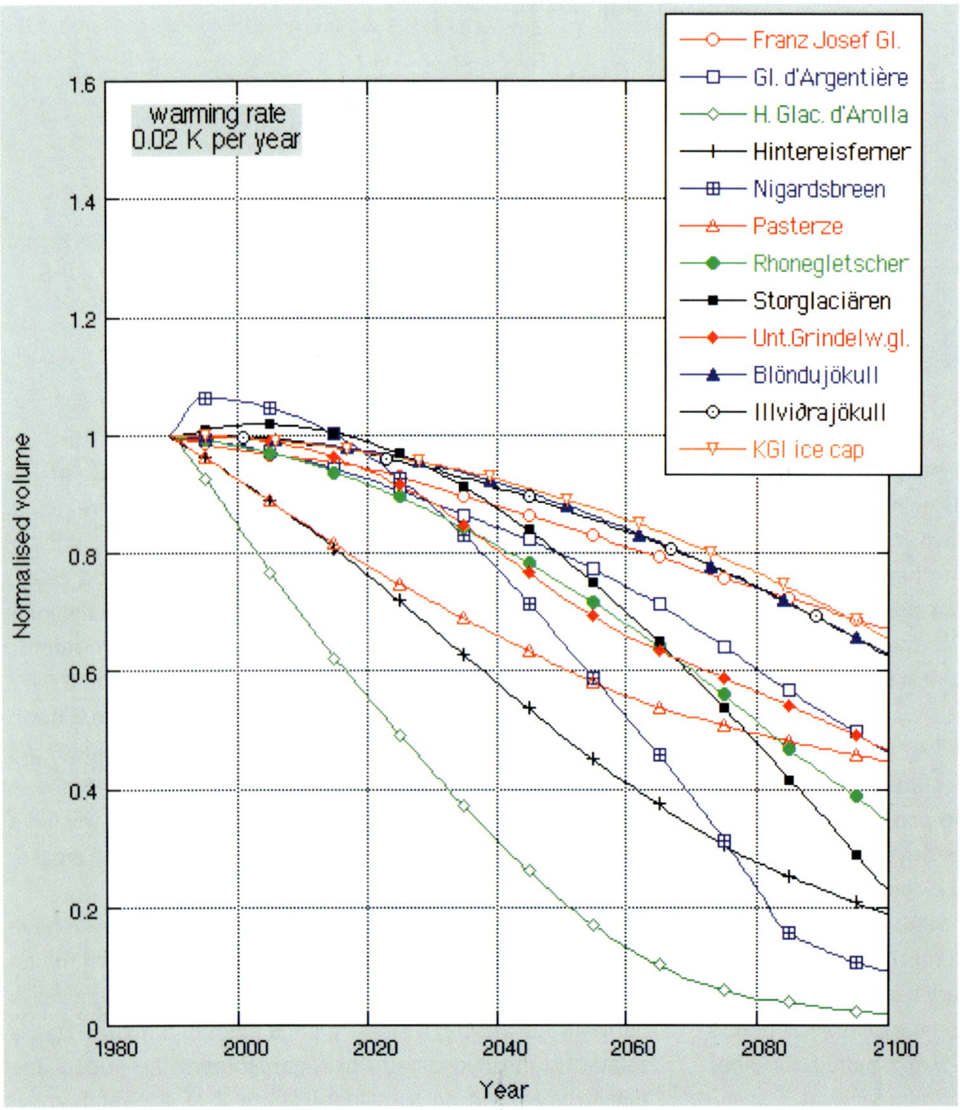

Figure 11.2. Results of the EISMINT-study for the case of a constant warming rate of 0.02 K a^{-1} from 1990 onwards. The curves give the volumes of the glaciers studied, normalised with the corresponding volumes for 1990 [2].

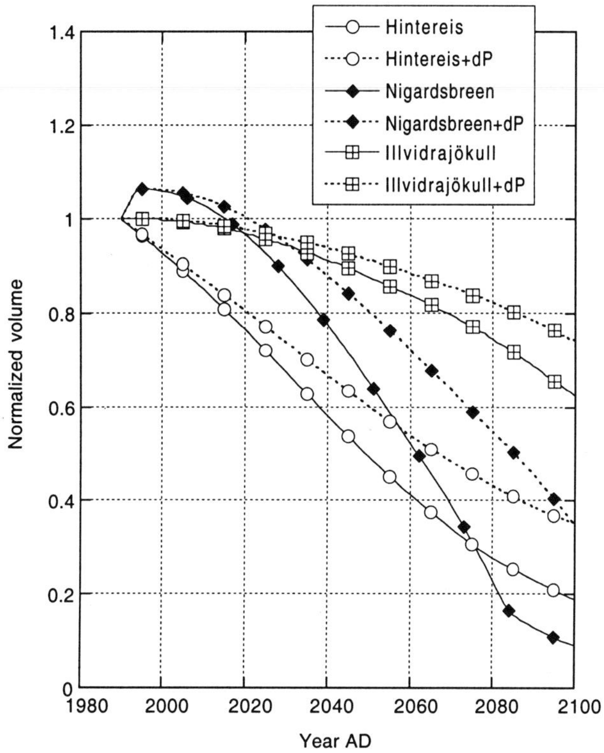

Figure 11.3. A comparison of the normalised volume for the 0.02-scenario and the 0.02^+-scenario, in which precipitation is assumed to increase by 10% per degree warming. Results are shown for three glaciers [2].

On the other hand, one can add together all volume changes and scale the resulting curve with the total 1990-volume. We refer to this curve as $\langle V \rangle_{sc}$. It is clear that the $\langle V \rangle_{sc}$ puts much more weight on the larger ice masses like the KGI ice cap, Illvidrajökull and Blondujökull. *Figure 11.4* shows $\langle V_{sc} \rangle$ and $\langle V \rangle_{sc}$ for the period 1990- 2100. As expected, there is a large difference between the curves. Unfortunately, area class distributions of glaciers are available for limited regions only (Haeberli et al., 1989), so it is impossible to validate either of the resulting curves. It is likely, however, that the curve representing the real situation lies somewhere in between the $\langle V_{sc} \rangle$-curve and the $\langle V \rangle_{sc}$-curve.

From a general point of view, the type of glacier that is obviously missing in the sample is the dry polar ice cap. Glaciers of this type are found in the Canadian Arctic, and on Svalbard, Frans-Josef Land, Novaya Zemlya and Severnaya Zemlya. There are also a number of dry ice caps on Greenland, which are not attached to the main ice sheet. The dry ice caps have a small climate sensitivity, but cover a large area. They also have long response times and it is likely that, if they were taken into account, they would shift the V_{sc}-curve upwards.

In the previous chapter we discussed how we could calculate the contribution of glaciers to sea-level change in the past. In this calculation the area covered by glaciers could not change in the course of time. We refer to such an approach as the fixed-geometry method (FG-method). The FG-method has also been used to make projections of sea-level change in the future, e.g. [5]. In the EISMINT-study the validity of the FG-method was tested using the sample of 12 glaciers and ice caps. The analysis is discussed briefly below.

In the FM-method, the volume of an individual glacier is computed from:

$$V(t) = V(t_0) - C_T A(t_0) \int_{t_0}^{t} T' \, dt \, . \qquad (11.2)$$

Here V(t) is the volume at time t, $V(t_0)$ is the initial volume, C_T is the sensitivity of the mean specific balance to annual temperature (section 4.7), $A(t_0)$ is the glacier area at time t_0, and T' is the annual temperature perturbation [note: in practice t is a discrete variable with a time-step of one year]. Now it is assumed that the temperature perturbation increases linearly with time (a is a constant):

$$T' = a\,(t - t_0) \, . \qquad (11.3)$$

In this case the integral can easily be evaluated and the normalised ice volume becomes:

$$\frac{V(t)}{V(t_0)} = 1 - \frac{a\, C_T A(t_0)\, (t - t_0)^2}{2\, V(t_0)} \, . \qquad (11.3)$$

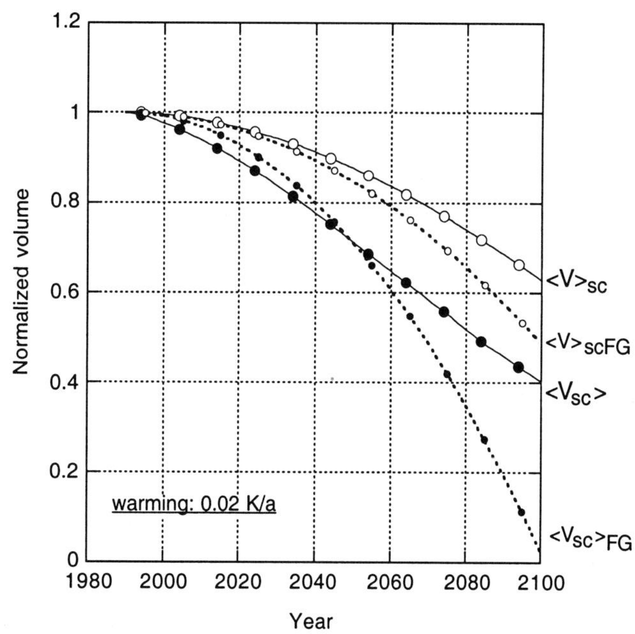

Figure 11.4. Normalised volume for the 0.02 K a^{-1} warming experiment. The curve labelled $\langle V \rangle_{SC}$ shows total volume scaled with the 1990 total volume. This curve is dominated by the ice caps. The curve labelled $\langle V_{SC} \rangle$ shows the mean of the scaled volume of the individual glaciers. In this case all glaciers in the sample have equal weight. Also shown are the associated results from a fixed-geometry approach, in which loss of ice volume is calculated from the static sensitivity (labelled FG).

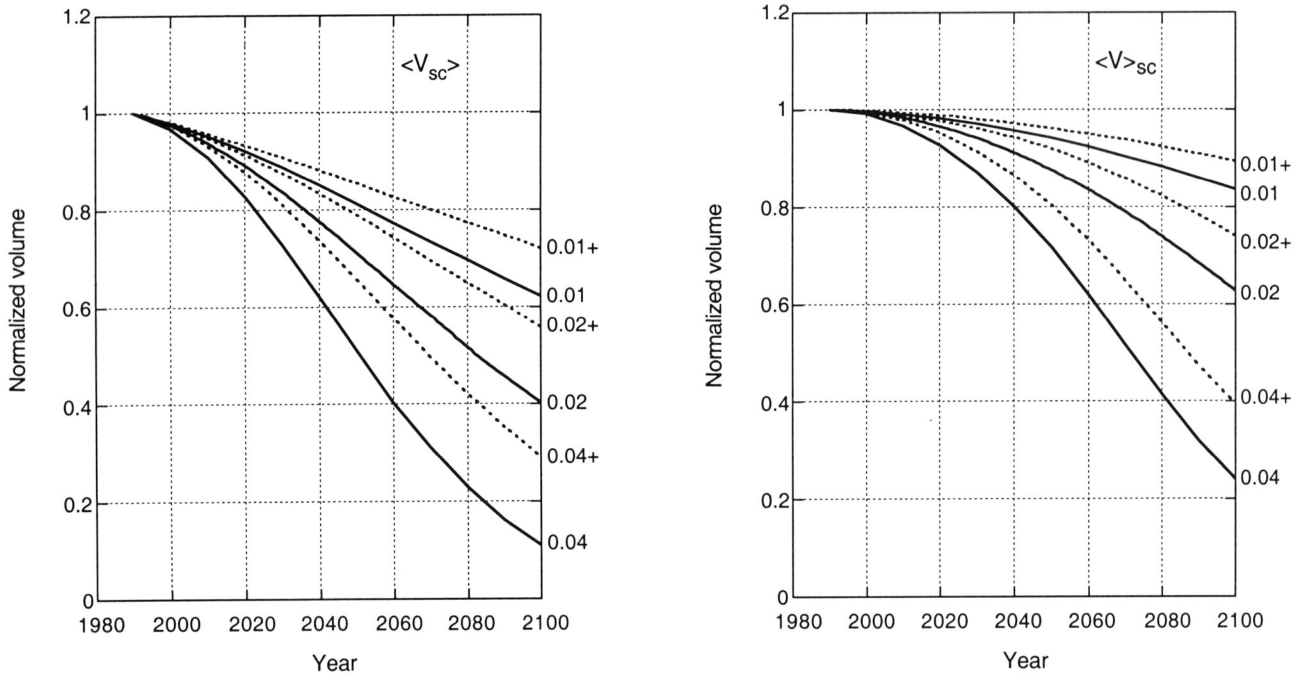

Figure 11.5. Scaled ice volume for all climate change scenarios used in the EISMINT-study. Labels on the curves give the imposed warming rates (+ refers to the case with an increase in precipitation of 10% per degree warming). Again the $<V>_{SC}$ curves are dominated by the ice caps.

Averaging now over the entire sample of glaciers (giving equal weights) yields:

$$<V_{sc}>_{FG} = \frac{1}{12}\sum_{i=1}^{12}\frac{V_i(t)}{V_i(t_0)} = 1 - \frac{a\,(t - t_0)^2}{24}\sum_{i=1}^{12}\frac{C_{T,i}\,A_i(t_0)}{V_i(t_0)} \quad . \qquad (11.4)$$

The index FG refers to fixed geometry, the index i is the glacier number. Similarly, to obtain a curve that can be compared to $<V>_{SC}$, we define:

$$<V>_{sc\,FG} = \frac{\sum_{i=1}^{12} V_i(t)}{\sum_{i=1}^{12} V_i(t_0)} = 1 - \frac{\frac{1}{2}a\,(t - t_0)^2 \sum_{i=1}^{12}\left\{C_{T,i}\,A_i(t_0)\right\}}{\sum_{i=1}^{12} V_i(t_0)} \quad . \qquad (11.5)$$

For the 0.02 experiment the quantities defined in eqs. (11.4) and (11.5) are also shown in *figure 11.4*. Apparently, after about 70 years the FG-method becomes less accurate and gives rise to significant errors. For the case of scaled total volume ($<V>_{SC}$), the FG-method always overestimates the loss of ice. In this case the large ice caps, which were supposed to be in equilibrium in 1990, dominate. However, for the case of mean scaled volume ($<V_{SC}>$), the situation is different. Apparently, the FG-method underestimates the loss of ice until about 2050. This is due to the fact that, on average, the glaciers in the sample were already retreating at the start of the integration. Therefore the effect of the dynamic calibration procedure is clearly visible here.

Finally, we will look at the results of the EISMINT-study for other climate scenarios.

In *figure 11.5* results for all climate scenarios used in the EISMINT-study are summarised. For a warming rate of 0.04 K a^{-1}, without a significant increase in precipitation, little ice would be left by the year 2100. On the other hand, if the warming rate were limited to 0.01 K a^{-1} with a substantial increase in precipitation, the loss of ice would be restricted to 10 to 20% of the 1990 volume.

11.3 The EISMINT-results in one equation

In this section we investigate whether it is possible to describe the results of the EISMINT-study in a single equation that describes the loss of ice volume for a given temperature scenario. We will concentrate on the curves for $<V>_{SC}$, and consider two models: a simple linear model and a model in which the time scale decreases with the magnitude of the temperature perturbation (as described in section 11.1). The corresponding equations are:

$$\frac{d\widetilde{V}}{dt} = -c_1\left\{\widetilde{V} - \exp\left[\frac{-(T - T_{ref})}{c_2}\right]\right\} , \qquad (11.6)$$

and

$$\frac{d\widetilde{V}}{dt} = -c_3\,(T - T_{ref})\left\{\widetilde{V} - \exp\left[\frac{-(T - T_{ref})}{c_4}\right]\right\} . \qquad (11.7)$$

Here $\tilde{V}$ is ice volume normalised with its value at $t = 0$. As discussed in section 11.1, the equilibrium ice volume decays exponentially with temperature with an e-folding temperature scale c_2.

The constants c_1, c_2, c_3, and c_4 can be found by minimising the root-mean-square difference between $\tilde{V}$ and the $<V>_{SC}$-curves as shown in *figure 11.5*. We assume that the initial state is balanced ($T_{ref} = 0$) and we optimise in such a way that one set of constants is obtained for all three scenarios (warming rates of 0.01, 0.02, 0.04 K a^{-1}). The results are shown in *figure 11.6*.

It appears to be impossible to make the linear model work well. The best result is obtained for $c_1 = 0.0286\ a^{-1}$ and $c_2 = 3.1\ K^{-1}$, and the corresponding root-mean-square difference is 0.0535. Clearly the model does not reproduce the EISMINT results in a satisfactory way. The nonlinear model performs much better. With $c_1 = 0.0066\ a^{-1}\ K^{-1}$ and $c_2 = 1.1\ K^{-1}$ the root-mean-square difference is only 0.0149. The e-folding temperature scale for ice volume (c_2) is quite small now, but care should be taken to make a physical interpretation. By fitting eq. (11.7) to the observed curves, c_2 becomes a control parameter and its value depends on the particular formulation of the volume response time. In view of the fact that less sensitive subpolar ice caps are not included in the set of glaciers considered in the EISMINT-study, the real value of c_1, representing the entire global state of glaciers and ice caps, is likely to be larger.

Why do the linear and the nonlinear models perform differently? The physical interpretation of the nonlinear model is not straightforward, unfortunately. Apparently good results are obtained when the characteristic time scale is smaller for larger temperature perturbations. Such an approach works well for a scenario of constantly increasing temperature, but it will be less accurate if a new equilibrium is approached, because then the time scale will become smaller again. The linear model fails because it is unable to reproduce the rapid increase in melt that occurs after some time due to the lowering of the ice surface. The acceleration of the mass loss cannot be captured with a single time scale. This result applies not only to the normalised volume of the entire sample, but also to the individual glaciers and ice caps.

Although the nonlinear model described by eq. (11.7) performs well for the given temperature scenarios, its use is limited to a period of 100 years. A model in which the volume response time depends on the difference between actual volume and equilibrium volume is physically more plausible. It can be described by the equation

$$\frac{d\tilde{V}}{dt} = -c_5 \left\{ \tilde{V} - \exp\left[\frac{-(T - T_{ref})}{c_6} \right] \right\}^m . \qquad (11.8)$$

The value of m would be equal to 2 if the response time were indeed proportional to the difference between volume and equilibrium volume for the actual temperature perturbation. However, m can be used as an additional control parameter. Unfortunately, it turned out to be impossible to find values for c_5, c_6 and m that give a reasonable fit to the EISMINT-curves. The results are in between those for eqs. (11.6) and (11.7). Therefore we conclude that a more sophisticated theory is needed to describe the behaviour of a set of glaciers. A scaling approach along the lines described by Bahr [6] may help, but it has still to be demonstrated that the results from such an approach match those of more explicit glacier models.

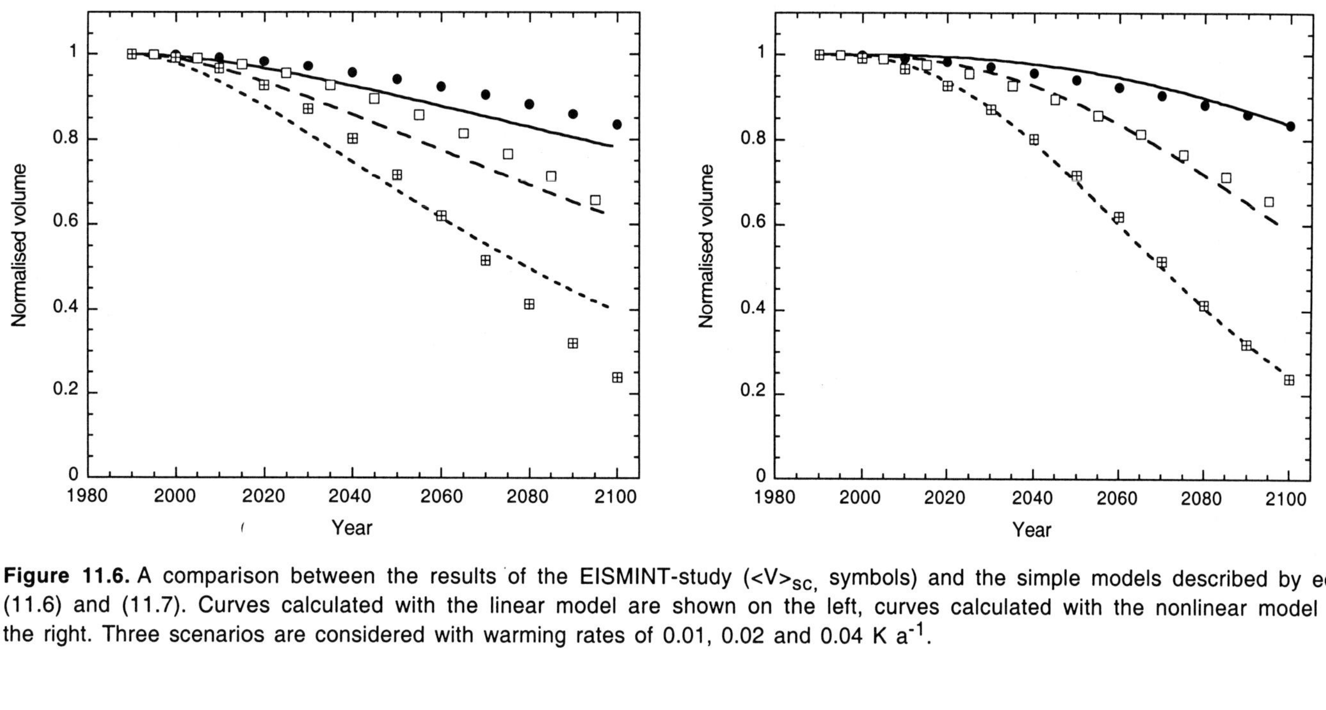

Figure 11.6. A comparison between the results of the EISMINT-study ($<V>_{SC}$, symbols) and the simple models described by eqs. (11.6) and (11.7). Curves calculated with the linear model are shown on the left, curves calculated with the nonlinear model on the right. Three scenarios are considered with warming rates of 0.01, 0.02 and 0.04 K a^{-1}.

11.4 Climate change and melt-water discharge from glaciers

In many parts of the world runoff from glaciers is used to fill hydropower reservoirs or to feed irrigation systems in dry valleys and plains. Obviously, enhanced greenhouse warming may have a significant effect on the water supply from glaciers. Here we are primarily interested in the behaviour of individual glaciers.

With regard to the total annual discharge, it may not make a large difference whether a catchment area has glaciers or not. If a glacier in a catchment area is in equilibrium with climate, its presence merely leads to a larger mean residence time of water [we ignore the more subtle effects that glacierisation may have on the micro-climate of a catchment area].

The seasonal dependence of runoff shows another picture, however [7]. This is best understood if we first consider a mountain catchment area in a warm climatic environment with a constant precipitation-evaporation rate through the year. If we neglect seasonal variations in the moisture-holding capacity of the soil, the discharge from the catchment area will be constant. When the climate is colder and seasonal snow occurs, part of the precipitation is stored in a snowpack in winter and released in spring and summer. This seasonality of the discharge may become even more pronounced when conditions are so cold that glaciers form in the catchment area. In that case melt-water is produced all the time until the next winter starts. The effect of glacierisation of a catchment area on the seasonal variation of the discharge is illustrated in *figure 11.7*. In the case of a small glacier, snow melt is still the most important process and the discharge peaks before the summer. With a large glacier the maximum discharge will occur later in the warm season.

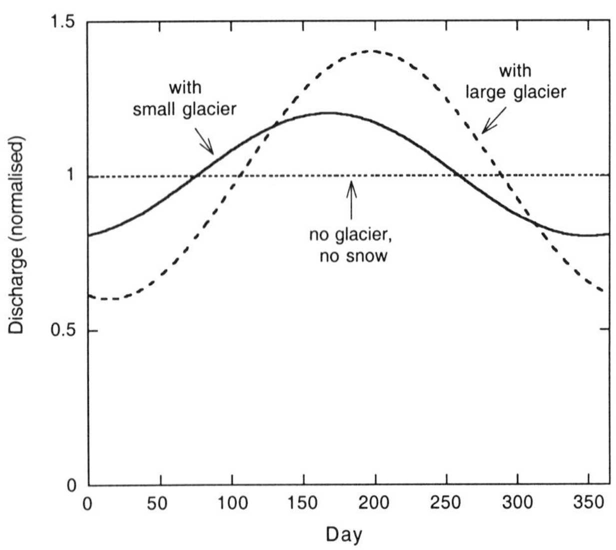

Figure 11.7. The effect of the degree of glacierisation in a catchment area on the seasonal cycle in discharge. The curves refer to the case in which the precipitation rate is constant through the year.

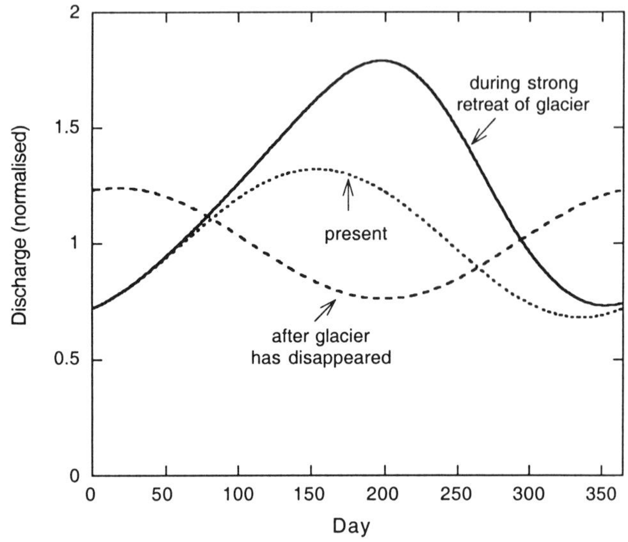

Figure 11.8. Discharge curves for a catchment basin with a large glacier that will disappear due to climatic warming. Precipitation has a seasonal cycle with a minimum in summer.

Melt water from glaciers becomes exceedingly important when the precipitation rate exhibits a seasonal cycle in such a way that the summer is the driest period. This is generally not the case in subpolar and mid-latitudes, but in many subtropical regions the amount of summer precipitation is very small. In some regions in central Asia, for instance, melt-water from glaciers is virtually the only source of water in summer.

Climatic warming may have severe consequences for the discharge in glacierised basins. A retreating glacier loses mass, which implies an increased discharge, mainly in summer. When most of the glacier has disappeared, the annual discharge may return to former levels, but the summer discharge will be less. This scenario is illustrated in *figure 11.8*. Schematic discharge curves are shown for a basin with a large glacier that disappears due to climatic warming. The precipitation has a marked seasonal cycle: in winter it is about twice as large as in summer. The curve labelled 'present' shows a discharge with a maximum in summer, associated with glacier melt. When the glacier retreats the total discharge increases; this is due almost entirely to enhanced melting in the summer and early autumn. By the time the glacier has disappeared conditions are so warm that the storage of water as snow has become less significant. Therefore the discharge curve now reflects the seasonal variation in precipitation with a minimum in summer. In the example shown in *figure 11.8* it has been assumed that the total annual precipitation does not change. Consequently, the curves labelled 'present' and 'after glacier has disappeared' represent approximately the same total annual discharge.

The situation sketched above may be rather extreme. The magnitude of the various effects depends on parameters like fraction of the catchment basin covered by glaciers, altitudinal range and hypsometry, annual cycle in air temperature, etc. The impact that retreating glaciers make on discharge will therefore vary strongly from place to place. The results of the EISMINT-study can be used to illustrate this. In *figure 9.11* the rate of change of normalised ice volume for three glaciers is plotted as a function of time for the 0.02-experiment. This volume change appears as additional discharge in the glacial river, so the curves in *figure 11.9* can be compared with the upper curve in *figure 11.8* (apart from the sign). Indeed, the curves for Nigardsbreen and Unterer Grindelwaldgletscher are of a similar shape: the discharge increases rapidly but then starts to decrease because the glaciers become smaller. For Franz-Josef Glacier the picture is different. Because the glacier is less sensitive the additional discharge increases steadily and then remains at a fairly constant level. If the integration were to be extended over another 100 years, the discharge would of course diminish.

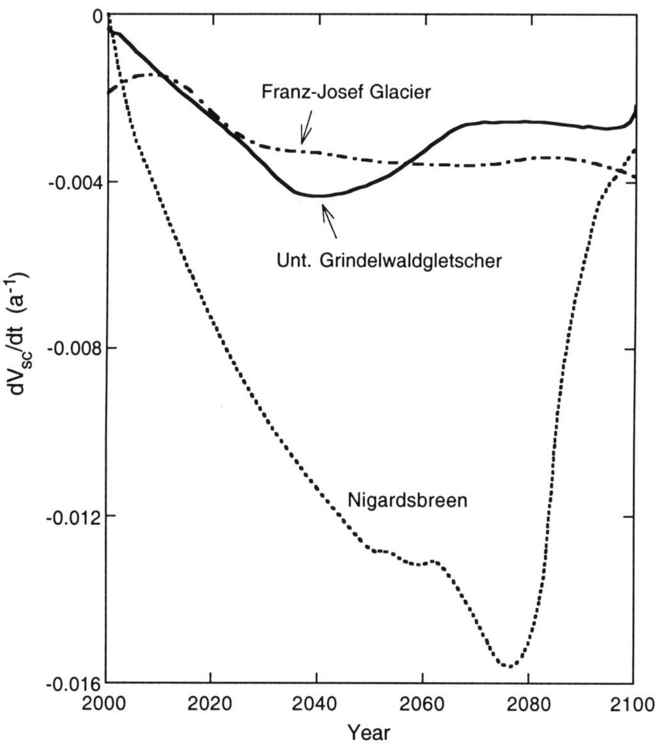

Figure 11.9. Annual change of volume of three glaciers, as computed in the EISMINT-study for a linear warming trend of 0.02 K a^{-1}.

References and further reading

1. J. Oerlemans (1989): A projection of future sea level. *Climatic Change* **15**, 151-174.

2. J. Oerlemans, B. Anderson, A. Hubbard, Ph. Huybrechts, T. Jóhannesson, W.H. Knap, M. Schmeits, A.P. Stroeven, R.S.W. van de Wal, J. Wallinga and Z. Zuo (1998): Modelling the response of glaciers to climate warming. *Climate Dynamics* **14**, 267-274.

3. W.H. Knap, J. Oerlemans and M. Cadée (1996): Climate sensitivity of the ice cap of King George Island, South Shetland Islands, Antarctica. *Annals of Glaciology* **23**, 154-159.

4. W. Haeberli, H. Bösch, K. Scherler, G. Østrem, C.C. Wallén (1989) *World Glacier Inventory. Status 1988.* IAHS(ICSI)-UEP-UNESCO.

5. J. de Wolde, P. Huybrechts, J. Oerlemans and R.S.W. van de Wal (1997): Projections of global mean sea level rise calculated with a 2D energy-balance climate model and dynamic ice sheet models. *Tellus* **49A**, 486-502.

6. D.B. Bahr (1997): Width and length scaling of glaciers. *Journal of Glaciology* **43**, 557-562.

   M. Dyurgerov and D.B. Bahr (1999): Correlations between glacier properties: finding appropriate parameters for global glacier monitoring. *Journal of Glaciology* **45**, 9-16.

   D.B. Bahr and M. Dyurgerov (1999): Characteristic mass-balance scaling with valley glacier-size. *Journal of Glaciology* **45**, 17-21.

7. G. Østrem (1973): Runoff forecasts for highly glacierized basins. In: *The role of snow and ice in hydrology*, IAHS Publ. no. 107, 1111-1132.

   H. Röthlisberger and H. Lang (1987): Glacial hydrology. In: *Glacio-fluvial sediment transfer* (eds. A.M. Gurnell, M.J. Clark), John Wiley and Sons, New York, 207-284.

   L.N. Braun and M. Aellen (1990): Modelling discharge of glacierized basins assisted by direct measurements of glacier mass balance. In: *Hydrology of mountain regions I*, IAHS Publ. no. 193, 99-106.

   H. Escher-Vetter and O. Reinwarth (1994): Two decades of runoff measurements (1974-1993) at the Pegelstation Vernagtbach / Oetztal Alps. *Zeitschrift für Gletscherkunde und Glazialgeologie* **30**, 53-98.

12. The future of glacier research

12.1 Introduction

Environmental science is being revolutionised by new technology. Computer capacity is increasing at an amazing rate, and instruments on board space platforms are sending down gigabytes of data every hour. But is glacier research benefiting from these developments? It is, of course, but rather slowly.

It is dangerous to write about possible future developments in a textbook, because the picture may change rapidly. Nevertheless, a selection of material is presented here that may serve as a source of inspiration for future studies.

The strategy used for observations and monitoring is changing. Traditional measurements of glacier geometry performed with geodetic methods and air photogrammetry are now complemented by satellite observations. The main advantage of the satellite input is the global coverage - not the accuracy of the measurements. Publications dealing with remote sensing in glaciology demonstrate that more work has been done on the larger polar ice masses than on valley glaciers. This is probably related to the fact that higher resolution is required for valley glaciers.

Optical sensors on board space platforms have not been used very much for glacier studies, except for mapping purposes (Landsat-TM). Nevertheless, ablation on glaciers is first of all a matter of absorption of solar radiation in the visible part of the spectrum. The evolution of the melt process in the course of the warm season is known to be strongly reflected in the evolution of the albedo pattern. As a consequence, mapping of the albedo pattern should provide information about a glacier's mass balance. This issue is discussed in section 12.2.

Using microwaves has a huge advantage over optical sensors: the signal is much less disturbed by conditions in the atmosphere. Passive microwaves are not useful for valley glacier studies, because the resolution (tens of kilometres) is too poor. However, radar from space platforms is an exciting new technique offering a resolution that is good enough for the study of valley glaciers. A few aspects will be described in section 12.3.

In recent years, there has been a revival of remote sensing of glaciers and ice sheets from aircraft. Traditionally planes have been used for air photogrammetry, but now laser altimetry combined with kinematic GPS provides a more efficient method for determining whether a glacier is thinning or thickening. Possibly, in the future validation studies of satellite products will make more extensive use of planes and helicopters equipped with the type of sensors that are also on board space platforms.

What therefore will be the role of conventional measurements at the glacier surface (stakes, snow pits, snow radar, weather stations, etc.)? In section 12.4 it will be argued that we cannot do without these measurements, but we need a better strategy to reduce their cost and increase their scientific merit.

The final section is devoted to modelling. If we want to be able to make statements about the future behaviour of glaciers, we need to aggregate data of different kinds. We cannot do this without a mathematical description of the processes we are looking at.

12.2 Using satellite data: optical sensors

Sensors operating in the visible part of the spectrum deliver images that are very appealing, because they show what we can also observe by eye: the intensity of reflected solar radiation. Because glaciers and ice caps generally have a higher albedo than the surroundings, optical sensors delineate their extent (although it can be difficult to identify the margin of glaciers with a heavy debris cover).

Figure 12.2. Landsat scene (Thematic Mapper band 2) of Breidamerkurjökull (see arrow in *figure 12.1*). The resolution is about 100 times better than the NOAA-AVHRR images. Icebergs can be seen in front of the glacier snout. Even the breaking of waves on the shore is resolved (white line marking the coast).

May 13th, 1996

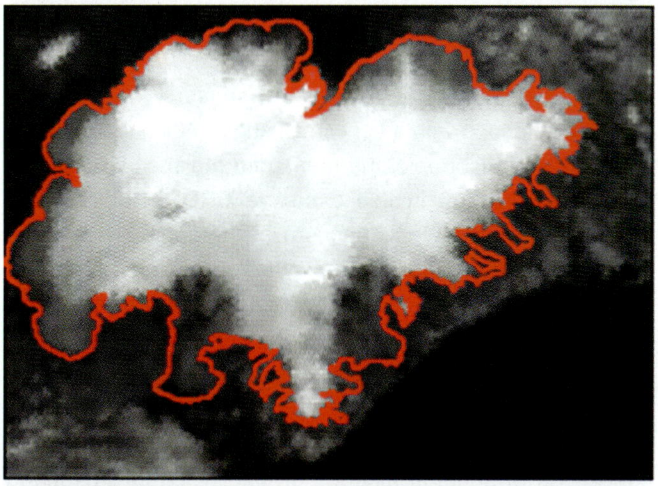

August 25th, 1996

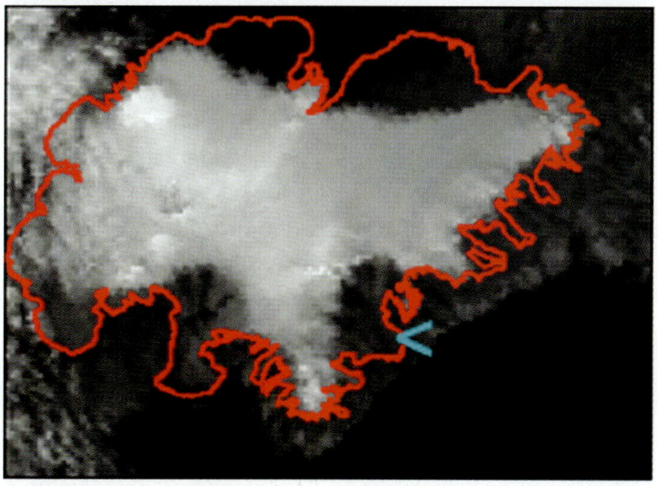

September 8th, 1996

Figure 12.1. Three NOAA-images (AVHRR-Channel 2) of Vatnajökull, Iceland (compare with *figure 3.14*). The pictures were made in the melting season of 1996. The red contour indicates the approximate margin of the ice cap and its outlet glaciers. The blue arrow points to Breidamerkurjökull, of which a Landsat-TM scene is shown in *figure 12.2*. In the upper picture high ground outside the ice-cap area is still covered with snow. In the lower picture there are clouds on the left. The blue arrow points to the region shown in *figure 12.2*. The area shown measures approximately 220 x 160 km.

The albedo provides information about the state of the glacier surface, because the albedo of bare ice is normally lower than that of firn and snow. Consequently, in most cases the transient snowline can be readily identified.

The polar orbiting weather satellites of NOAA (USA) have delivered a wealth of data in visible and infrared channels. Unfortunately, the spatial resolution of the images is poor (~ 1 to 3 km, depending on the view angle). However, images are available for a long period of time (several decades), and the temporal resolution is good, implying that a useful scene is available for every day with little or no cloud cover.

In *figure 12.1* three AVHRR-images of Vatnajökull are shown. This series nicely demonstrates the progress of the melt process during the summer as reflected by the decreasing area of high albedo and clearly reveals the upward migration of the snowline. On the last image the transition from firn to (very dirty) ice is very sharp. The upper picture also reveals that Breidamerkurjökull, the large outlet that flows into the Atlantic (arrow in *figure 12.1*), was already free of snow by the middle of May.

Images of the Landsat thematic mapper are in sharp contrast to those of the AVHRR. They have high spatial resolution (~ 20 m), but poor temporal resolution because the typical repeat time (the time that elapses before the satellite passes over the target again) is a few weeks. Since many glaciers are located in regions with high mean cloudiness, one is lucky if one obtains a few good scenes in a single summer. For the earlier Landsat missions the availability of high-quality images has been documented [1].

A landsat image of Breidamerkurjökull is shown in *figure 12.2*. Many small-scale features can be seen: medial moraines, the calving lakes, icebergs, waves breaking on the beach, and, in particular, black tephra layers that are melting out in the ablation zone.

From these examples it is clear that NOAA-AVHRR images are not very useful for the study of valley glaciers because the resolution is not good enough. The archive of Landsat images, however, has certainly not been fully exploited (a possible reason being the high cost involved in acquiring scenes). With regard to the study of valley glaciers, Landsat scenes have been used mainly to delineate the glacier margin and classify the type of surface [2], and to determine the surface albedo [3]. All of the applications have their problems and further research is needed to improve the algorithms involved.

As noted in the introduction, the evolution of the surface albedo is strongly related to the mass-balance conditions at the glacier surface. Normally one can see the transition from a snow-covered surface to an ice surface (the snowline). The speed at which the snowline moves up-glacier is determined primarily by the amount of energy available for the melt process. However, the amount of snow that has fallen in winter also plays a role.

If there is a large amount of precipitation in winter, the altitudinal gradient of the snow depth at the beginning of the melt season will be larger than normal. This leads to slower migration of the snowline.

Is it possible to relate the mean specific balance of a glacier or a small ice cap to the evolution of the albedo pattern? We can define a quantity A_P, the potential absorption of solar radiation, as follows:

$$A_P = \frac{1}{S_{tot}} \int_{glacier} Q_E (1-\alpha)\, dx\, dy \quad . \qquad (12.1)$$

Here Q_E is the extra-terrestrial irradiance as discussed in Chapter 3, α is the albedo and x,y are the horizontal coordinates. The integral is taken over the entire glacier surface, which has a total area S_{tot}. For glaciers and small ice caps the variation of Q_E with latitude is insignificant, so Q_E can be taken as a constant for a given date and time. Next A_P is integrated over a balance year:

$$\widetilde{A_P} = \int_{year} A_P\, dt \quad . \qquad (12.2)$$

The quantity $\widetilde{A_P}$ has unit J m^{-2}. Its value will always be much larger than the actual amount of energy absorbed at the surface, because the solar radiation reaching the surface is much smaller than Q_E (Chapter 3).

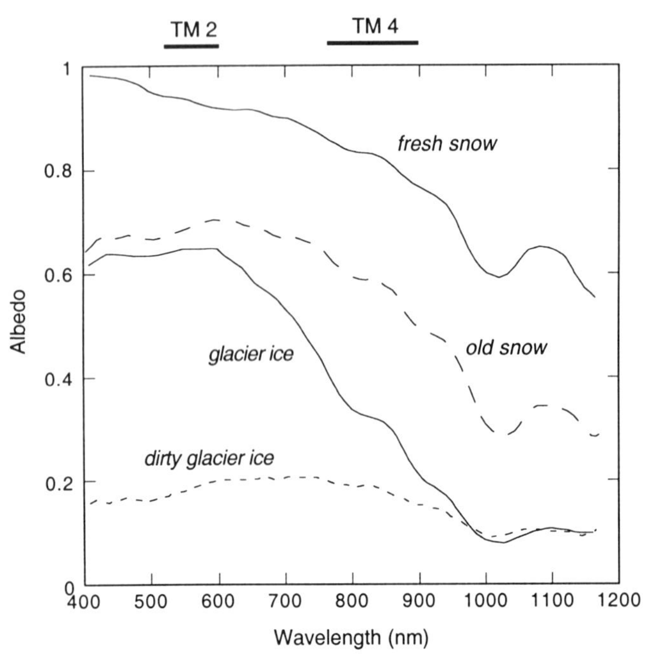

Figure 12.3. Spectral reflectance curves for ice and snow, as given by Hall et al. [3]. Note that the curves give the characteristic patterns only. The spectral width of the Landsat Thematic Mapper bands 2 (visible) and 4 (near-infrared) are shown at the top.

Because the absorption of solar radiation plays such an important role in the melt process, one might expect the mean specific balance and $\widetilde{A_P}$ to be correlated. Surprisingly, to date no studies along these lines have been published [note that this approach differs fundamentally from the approach in which the equilibrium line is determined by detecting the highest position of the transient snowline]. A possible reason is that the calculation of the albedo from satellite images is not a trivial matter.

The determination of the surface albedo involves a few complicated steps. In the first place, a conversion has to be made from band albedos (as measured by satellites) to spectrally integrated albedos (relevant to the surface energy budget, sometimes referred to as broadband albedo). Great care has to be taken here, because the spectral distribution of the reflected solar radiation depends on the nature of the surface (see *figure 12.3*).

In the second place, the satellite sensors measure the radiance in a cone with a very small angle. When the reflected radiation is not isotropic, the measured radiance will depend on the view angle. In this case the BRDF (Bi-directional Reflectance Distribution Function) of the surface has to be known. So the calculation of the albedo requires the converson of band albedos to a spectrally integrated albedo, as well as the conversion from a narrow-beam reflectance to a hemispherically integrated reflectance. We will now discuss these steps in more detail.

Knap et al. [4] have carried out an extensive study on the conversion of spectral albedos of glacier ice to broadband albedo for Landsat-TM images. LGRs (Landsat Global Radiometers) were designed to measure the surface reflectance in the Landsat TM bands 2 and 4 (see *figure 12.3* for the location of these bands in the spectrum). These instruments were used during two periods (April - September 1996 and February - April 1997) on the Morteratschgletscher. This work produced a large data set consisting of simultaneous observations of α (the broadband albedo), α_2 and α_4 (albedos in the TM bands 2 and 4). An empirical approach was used to relate α to the band albedos :

$$\alpha = a\,\alpha_2 + b\,\alpha_2^2 + c\,\alpha_4 + d\,\alpha_4^2 \quad . \qquad (12.3)$$

The coefficients a, b, c and d are determined by regression analysis. Knap et al. [4] analysed the data in various ways, making a distinction between the different types of surfaces. However, it turned out that a good fit between the measured and the parameterised albedo could be obtained without specifying the type of surface a priori, which is a great advantage. The best fit reads:

$$\alpha = 0.726\,\alpha_2 - 0.322\,\alpha_2^2 - 0.051\,\alpha_4 + 0.581\,\alpha_4^2 \quad . \qquad (12.4)$$

This fit is shown in *figure 12.4*, together with the data points. The correlation coefficient is very high ($r^2 = 0.998$).

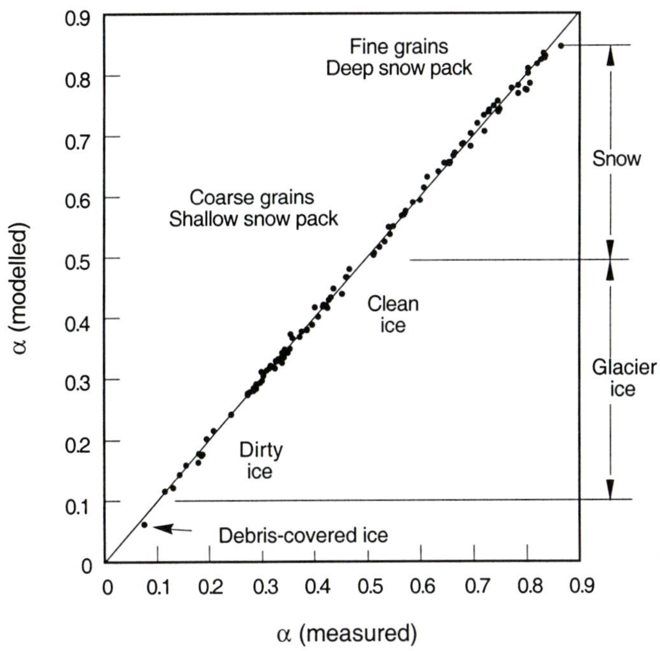

Figure 12.4. The relation between measured albedo and albedo calculated from TM band 2 and 4 radiances with eq. (12.4) [4].

Because the TM band 2 saturates over snow when the global radiation is large, it is useful to have an alternative relation from the same data set in which α_2 does not appear. This relation reads:

$$\alpha = 0.782\,\alpha_4 + 0.148\,\alpha_4^2\ . \qquad (12.5)$$

Again, the fit is quite good ($r^2 = 0.983$).

The results obtained by Knap et al. are in broad agreement with earlier studies in which more limited data sets were used [3]. However, far more data from glaciers and ice sheets are needed to find out how wide the applicability of eq. (12.4) really is.

Next we discuss the Bi-directional Reflectance Distribution Function (BRDF), which is needed to convert radiance (unit: W m^{-2} sr^{-1}), as measured by a satellite, to the reflected radiation integrated over the upper hemisphere (the irradiance; unit: W m^{-2}). The bi-directional reflectance $r(\theta_v,\phi)$ is defined as

$$r(\theta_v,\phi) = \frac{\pi\, R(\theta_v,\phi)}{I}\ . \qquad (12.6)$$

Here θ_v is the view zenith angle, ϕ is the relative azimuth (relative to the solar principal plane, i.e. the sun is at $\phi = 0$), I is the hemispheric flux (irradiance) of incoming radiation, and R is the radiance in the direction (θ_v,ϕ). Note that $\phi = 180^\circ$ corresponds to forward scattering and $\phi = 0^\circ$ corresponds to backward scattering. For an isotropically reflecting surface (normally referred to as a Lambertian surface), $r(\theta_v,\phi)$ is independent of the view angle and equals the albedo. The anisotropic reflection factor $f(\theta_v,\phi)$ can now be defined as

$$f(\theta_v,\phi) = \frac{r(\theta_v,\phi)}{\alpha}\ . \qquad (12.7)$$

It is important to realise that $f(\theta_v,\phi)$ depends on the solar zenith angle and on the wavelength. The anisotropy of the reflected radiation tends to be larger when the sun is lower. This can be observed by eye when the sun is very low: a glacier surface looks much brighter when one looks in the direction of the sun than when one looks away from the sun.

Figure 12.5. Studying the albedo of firn in the accumulation zone of a glacier. The instrument on the left measures the incoming radiation in the full spectrum and in Landsat TM bands 2 and 4. The instrument on the right measures the reflected radiation in a particular direction in Landsat TM bands 2 and 4.
The construction is designed in such a way that the radiance can be measured for many view angles in a short time.
Photograph by W. Greuell [6].

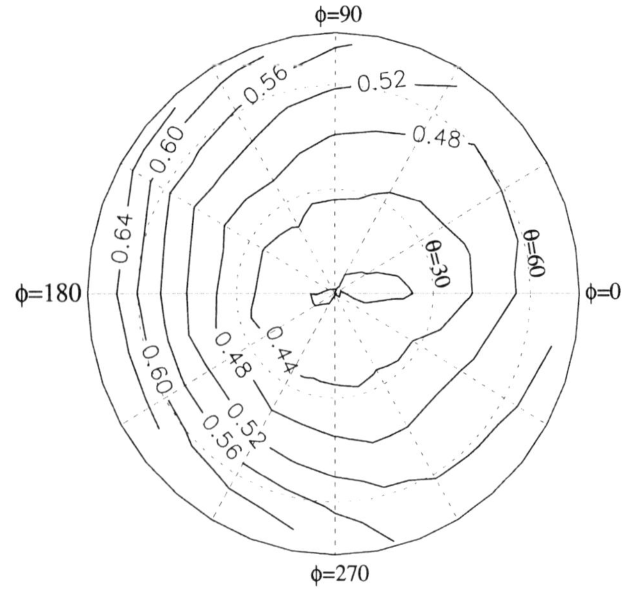

Figure 12.6. An example of a BRDF measured with the equipment shown in *figure 12.5*. The isolines give the anisotropic reflection factor f (labels on the curves). The measurements were made in TM band 4 over relatively dark ice, for a solar zenith angle of 50°. In this case the TM band 4 albedo was 0.5.

It appears that the reflection characteristics of snow have been studied more extensively than those of melting glacier ice [5]. In general, the degree of anisotropy is larger for an ice surface than for a snow surface.

In recent years BRDFs over melting glacier ice have been measured [6]. Here we describe some results obtained by Greuell and De Ruyter de Wildt. *Figure 12.5* shows the experimental set-up they used. The construction allows measurements to be made for a wide range of solar zenith angles (26° to 75°). An example of a BRDF, based on about 190 measured radiances for different values of $f(\theta_v,\phi)$, is shown in *figure 12.6.* The pattern is characterised by a minimum value of f in the nadir direction and relatively large values in the forward limb. It is clear that the degree of anisotropy in the reflection patterns is quite large. Greuell and De Ruyter de Wildt measured a large number of BRDFs and concluded that the anisotropy becomes larger for increasing wavelength and increasing solar zenith angle. They also found a relation between the anisotropy and the albedo: for smaller albedos the anisotropy tends to be larger. Greuell and De Ruyter de Wildt also developed parameterisations for $f(\theta_v,\phi)$, but these will not be discussed here.

It should be noted that the view zenith angle of the Landsat Thematic Mapper does not exceed 8°. In general, the corresponding variation of f is small. However, in the case of glaciers the surface slope of larger areas is typically 5° to 25°, which implies that values of f may be quite different from those over horizontal terrain. Therefore it is advisable to use a digital elevation model to determine the appropriate view angle for individual pixels in Landsat-TM scenes. In such a procedure accurate geolocation of the Landsat scenes is essential.

Two other aspects of working with Landsat-TM and NOAA-AVHRR images have not been discussed, namely, sensor calibration and atmospheric effects.

Neither for the NOAA weather satellites nor for the Landsat-TM sensors is there onboard calibration; the only basic information available is from pre-launch experiments [7]. Attempts have been made, however, to do in-flight calibration by looking at very bright and dark surfaces seen by the satellite [8]. Nevertheless, calibration of the sensors remains a sensitive issue.

Atmospheric effects can be handled with radiative transfer models. Accurate models exist; here the major source of uncertainty is the composition of the atmosphere. For the study of surface albedo one chooses clear-sky scenes, of course, but aerosol layers can still have a significant effect on the attenuation of incoming and reflected solar radiation [9].

All in all, the existing data sets of satellite measurements in the visible and near-infrared channels form a useful basis for the study of glacier mass balance, but more research into the nature of the reflection process at glacier surfaces is needed.

12.2 Using satellite data: microwaves

Large datasets are available from microwave radiometers that have been carried by a number of satellites. Due to the large wavelength (1 mm to 1 m) and the limited size of antennas on space platforms, the spatial resolution is always low (typically 10 km or more). Passive microwave radiometer data can be adequate for the study of large ice sheets, but they are of little value for glaciers and small ice caps.

Active microwave signals offer much more, because synthetic aperture techniques (SAR) allow a much higher resolution (typically 10 to 50 m). The backscatter depends on the dielectrical properties of the surface, and on the geometrical characteristics. Because microwaves penetrate into ice and snow, the signal returned from the surface to the satellite contains information about the structure of the upper layers - not just the surface. Although there are many theories about how microwaves interact with an ice or snow surface, a straightforward interpretation in terms of quantities that are relevant for the mapping and/or modelling of glacier mass balance appears to be difficult. It has been suggested that, under certain conditions, the upward migration of the melt line on a glacier can be traced with SAR images [10]. However, interpretation is not easy and the datasets are limited in time, which makes SAR less attractive for monitoring glacier mass balance. Modern instruments provide multi-channel data and also look at different polarisations. In the future, these will certainly make it easier to characterise surfaces with a

complex structure. Nevertheless, a great deal of research is needed to find out if such instruments are really useful for determining glacier mass balance from space.
A relatively new technique with interesting applications in glaciology is SAR interferometery (InSAR) [11]. Under certain conditions fields of surface velocity and topography can be determined with InSAR. Here we discuss the principle briefly and have a look at two examples. The discussion below is to a large extent based on the lectures given by H. Rott (University of Innsbruck) at the 1997 EISMINT Summer School.
We start by looking at how a radar image line is actually built up. An across-track image line is the result of time differences in the signal reception. As illustrated in *figure 12.7*, any undulation in the terrain will lead to distortion of the image. Slopes oriented towards the radar lead to compression of the image ('foreshortening'). In the case of steep slopes the signal from the mountain top may be received earlier than the signal from the bottom ('layover'). The signal in a layover region is difficult to interpret, because it contains backscatter from more than one slope. For large values of the view angle, shadows appear in the image. However, the view angles used for spaceborne SAR are normally sufficiently small to prevent the occurrence of large shadows. In the SAR technique the phase and the magnitude of the backscattered signal Λ is used for further analysis. Symbolically in complex notation:

$$\Lambda = \mathrm{Re}(\Lambda) + i\,\mathrm{Im}(\Lambda) \quad . \qquad (12.8)$$

A conventional SAR image is a magnitude image in which the backscatter coefficient is mapped. Interferometry makes use of the phase information.
The most widely used type of interferometery is repeat pass interferometry. For this technique two images taken at different times from similar orbits are used. Pixelwise multiplication of the complex signals from two SAR scenes yields the interferogram described by the phase function ϕ (* denotes the complex conjugate):

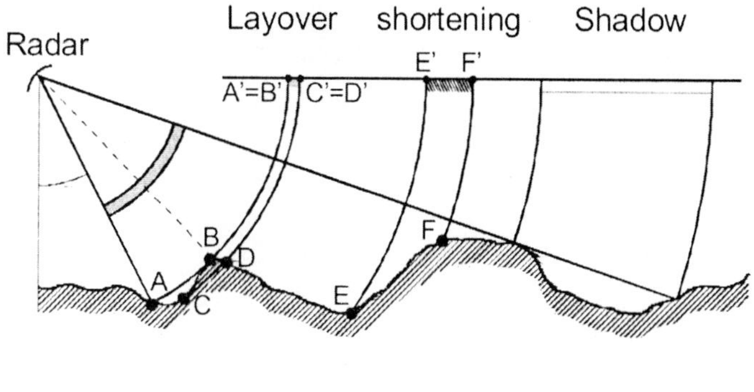

Figure 12.7. Illustration of how undulating terrain leads to distortion of the radar image in the across-track direction. From H. Rott (lecture notes for the 1997 EISMINT Summer School).

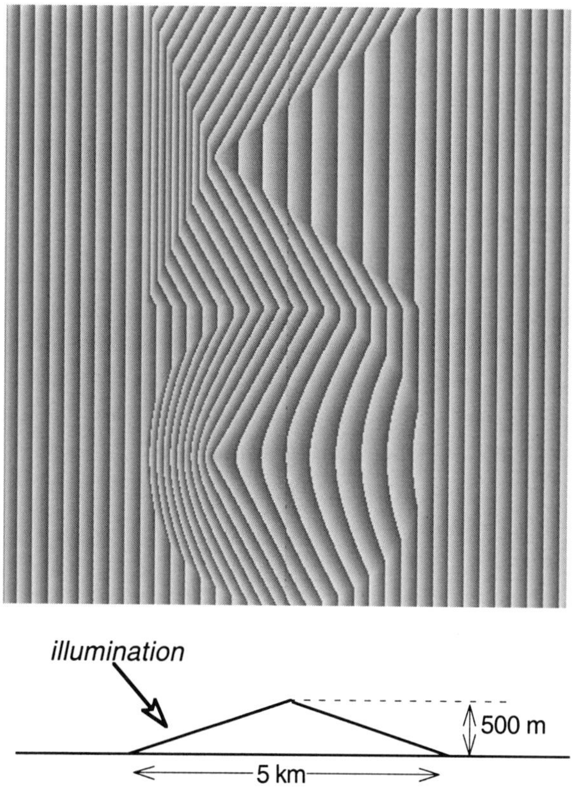

Figure 12.8. Interferogram for a pyramid (upper half) and a cone (lower half) calculated with ERS-SAR image parameters (baseline 100 m). From H. Rott (lecture notes for the 1997 EISMINT Summer School).

$$\phi = \arctan\left[\frac{\mathrm{Im}\,(\Lambda_1 \overset{*}{\Lambda}_2)}{\mathrm{Re}\,(\Lambda_1 \overset{*}{\Lambda}_2)}\right] \quad , \qquad (12.9)$$

An interferogram is only meaningful when the phase at the pixel scale is preserved. The degree of coherence γ_c is defined as

$$\gamma_c = \frac{E\{\Lambda_1 \overset{*}{\Lambda}_2\}}{\sqrt{E\{\Lambda_1\}^2 E\{\overset{*}{\Lambda}_2\}^2}} \quad . \qquad (12.10)$$

Here { } denotes ensemble averaging.
Many factors lead to decorrelation: noise in the data, geometric effects (differences in imaging geometry between the scenes), atmospheric effects, changes in the back-scattering properties of the surface (due to melting or snowfall, for example). For most satellites equipped with radar the time between two passes is of the order of a month, which is too long (the resulting value of γ_c is very low). Useful interferograms are obtained with scenes that are acquired with only one or two days time difference. This was for instance achieved with the ERS-1/ERS-2 Tandem Mission (August 1995 - May 1996), in which the orbits of the two satellites were such that the same area of the earth surface was imaged twice in a 24-hour time interval.

A simple example of an interferogram is shown in *figure 12.8*. In general the patterns are complex and the calculation of a topographic map from an interferogram is a complicated procedure.
As can be seen in *figure 12.8*, in the SAR imaging process the measured phase is known only modulo 2π. This is termed the wrapped phase, which does not allow the determination of unique solutions. The calculation of the topography requires the absolute geometric phase, so the sequence of fringes has to be determined (this process is called phase-unwrapping).
The variation in absolute phase $\Delta\phi_a$ across the image swath is made up of two contributions, namely, the flat earth contribution and the topographic contribution:

$$\Delta\phi_a = \Delta\phi_{flat} + \Delta\phi_{topo} \quad . \tag{12.11}$$

We will not discuss all the steps that have to be taken to derive a digital elevation model from two complex SAR images. It is clear that at least one, but preferably two or three accurate reference points are needed to anchor the SAR product and arrive at a reliable map in a geodetic reference system. Here precise orbit data are also a prerequisite.

If the target moves over a sufficiently large distance during the acquisition of two scenes the absolute phase difference also contains a motion component:

$$\Delta\phi_a = \Delta\phi_{flat} + \Delta\phi_{topo} + \Delta\phi_{motion} \quad . \tag{12.12}$$

It should be realised that this motion component results from the horizontal and vertical velocity of the target.
In the ideal situation where the orbits from which the two images were acquired are identical, $\Delta\phi_{flat}$ and $\Delta\phi_{topo}$ are zero and the motion field would follow directly. However, orbits are never exactly the same and more information is needed. By using two interferograms (differential interferometry) it is possible to separate the contributions from topography and motion (first the flat earth contribution should be eliminated; this is possible if the imaging geometry is known exactly).
An example of an interferogram based on two SAR images acquired by the Space Shuttle Endeavour in 1994 is shown in *figure 12.9* (Moreno Glacier, Southern Patagonia). In the areas where the fringes are wide the interferogram reflects the topography (one fringe corresponds to an altitude difference of 550 m).

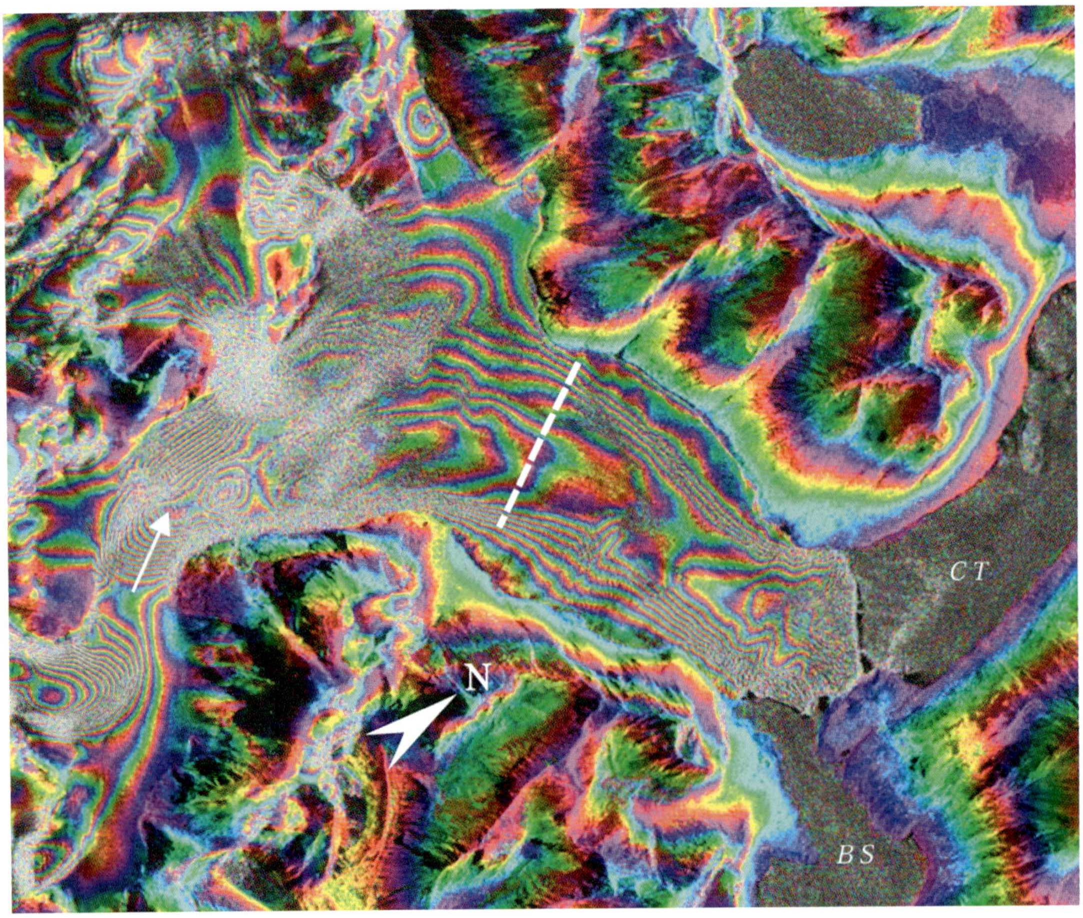

Figure 12.9. Interferogram of the Moreno Glacier, Southern Patagonian Icefield, obtained from the Spaceborne Imaging Radar-C/X-Band SAR that was flown on the Space Shuttle Endeavour (October 1994). The glacier front (lower right) calves into the Canal de los Témpanos (CT) and the Brazo Sur (BS). From Rott et al. [12].

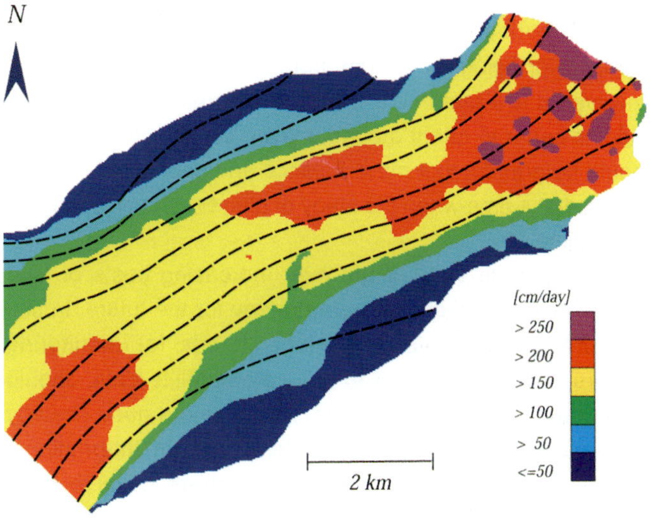

Figure 12.10. Ice velocity on the lower part of the Moreno Glacier deduced from SAR interferometry. From Rott et al. [12].

The glacier is characterised by much narrower fringes. Here the contribution from the motion field is much larger than from the topography. In some areas higher up the glacier the coherence is lost ($\gamma_c \approx 0$). The causes are probably too high velocities and/or rapid change of the surface texture due to melting or snowfall.

For the lower part of the glacier Rott et al. were able to deduce a map of the horizontal ice velocity (*figure 12.10*). The patterns shows a clear velocity maximum in the central part of the glacier. As we expect, velocities increase towards the calving front, where at some places values of more than 250 cm day^{-1} are found.

The method of differential interferometry described above yields only velocities in the plane perpendicular to the orbit. In order to obtain a complete velocity field one needs interferograms from two crossing orbits (i.e. at least four SAR images). Therefore the complete motion field of a glacier or ice cap can only be deduced in exceptional cases.

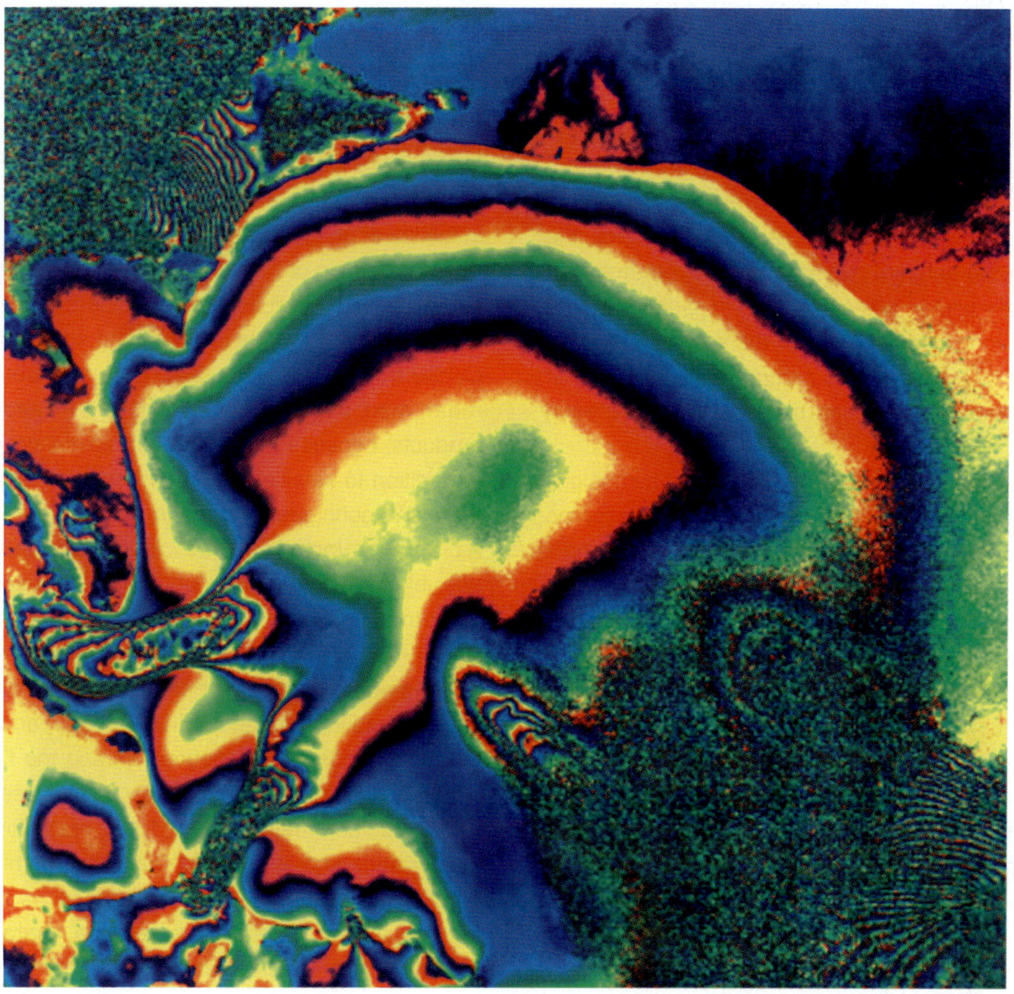

Figure 12.11. A mixed motion/topography interferogram of the Academy of Science Ice Cap, Severnaya Zemlya, obtained from two ERS-SAR images (December 1995). In the upper right and lower left part of the picture the coherence is very small. The smooth topography of the ice cap stands out, in contrast to a number of fast flowing outlets where the fringes are closely spaced. Provided by J. Dowdeswell, Bristol Glaciology Centre, U.K.

Another fine example (*figure 12.11*) is the interferogram of the Academy of Sciences Ice Cap, located in the Russian Arctic (Severnaya Zemlya). The two SAR images were acquired on the 2nd and 3rd of December 1995 during the ERS1/2 Tandem Mission. The interferogram is a mixed motion/topography interferogram. In some parts of the image the coherence is lost, most probably due to snowfall in the period between the images. The smooth topography of the main ice cap appears very nicely. Outlet glaciers where the motion component dominates can easily be identified.
The examples discussed above demonstrate the enormous potential of inSAR. In spite of the comparatively complex technique and the restricted availability of images (relative to the archives of visible and passive microwave scenes) inSAR allows the study of the dynamics of glaciers and ice caps of which very little is known. To what extent inSAR can be used for the actual monitoring of ice bodies remains an open question, but snapshots from topography and motion fields can be obtained in unprecedented detail.

12.4 The role of in situ measurements

Satellite data provide good spatial coverage. However, the accuracy of satellite products is generally not very good, because our knowledge of how electromagnetic waves interact with the glacier surface is limited. Furthermore, the calibration of satellite sensors may not be as accurate as we would wish, in particular when we use sensors that are operational for a long period. Many of the errors in satellite products are of a quasi-systematic nature, which implies that suitable in situ measurements can be used to reduce such errors.
It is probably fair to say that most of the measuring campaigns on glaciers have not been designed to deliver data sets for the calibration of satellite data (although some good attempts have been made [3]). More campaigns are certainly needed to test satellite products. If such campaigns are well designed they will greatly enhance the value of satellite measurements and will be cost-effective.
In situ measurements of glaciological and meteorological quantities are also needed because we want to understand the processes involved. Satellite data may tell us, for example, on how many days in a year more than 50% of a glacier experienced melting at the surface. However, even if we were to have this information over several years, it would still be difficult to find out how we could use this knowledge to estimate the climate sensitivity of this glacier's mass balance. Satellite data are a valuable tool in monitoring studies, but currently cannot replace in situ measurements.
Measurements on glaciers have not been set up as part of a global programme. They were designed mainly to serve the needs of local interested institutions like power companies or road authorities. Glaciological observations are often criticised for being too ad hoc from a scientific point of view. However, it is a fact that a wealth of data has been gathered and it is better to turn it around: maybe the scientists of today should be grateful that mass balance and geodetic data were collected at a time when the study of glacier dynamics was not a real issue.
Nevertheless, in view of the ever increasing costs of field work, we have to think about a strategy for the future. In situ measurements should serve local needs for monitoring glaciers, should be useful to calibrate satellite data, should support model development, etc. Developing a good strategy will require a great deal of thinking and an international approach. Fortunately, some new initiatives have recently been taken.
With regard to mass balance and climate it appears desirable to combine mass-balance measurements with meteorological observations on or very near the glacier. Ideally, a meteorological station on a glacier should measure air temperature, humidity, wind speed and direction, the four components of the radiation balance (Chapter 3), temperature of the snow pack, snow accumulation and ice melt. The resulting data set makes a variety of studies possible. The location for a weather station should be chosen with care. In autumn and winter icing is a major problem, especially when stations are located on or close to a dome where persistent clouds may occur in combination with strong winds. Stations in the lower parts of ablation zones are not so vulnerable.
The exploitation of snow radars is an interesting development, because such devices may considerably reduce the errors that are normally made in estimating snow accumulation on a glacier from a few pit studies. Data sets obtained with snow radar also allow comparisons to be made with satellite products, because they give good spatial coverage.
In addition to satellite measurements and in situ observations there are techniques that make use of planes and helicopters. Air photogrammetry is the classic example, but it is also possible to fly instruments of the type that are used on space platforms.
As an example, *figure 2.12* shows results from a helicopter flight over Breidamerkurjökull (*figure 12.2*). In this case reflected radiation in the AVHRR1 (visible) and AVHRR2 (near-infrared) bands was measured. During the flight (at very low altitude), the glacier tongue was first crossed, and then the major medial moraine (*figure 12.2*) was followed down to the glacier edge and onto the tundra. The incoming radiation was measured simultaneously at the surface.
First of all the medial moraines are characterised by low albedo in the visible and near-infrared bands. The ice in between has a relatively low and variable albedo, but it is clear that the near-infrared albedo is systematically lower. On the tundra, well covered by vegetation, the opposite is the case: here the near infra-red signal is significantly stronger.

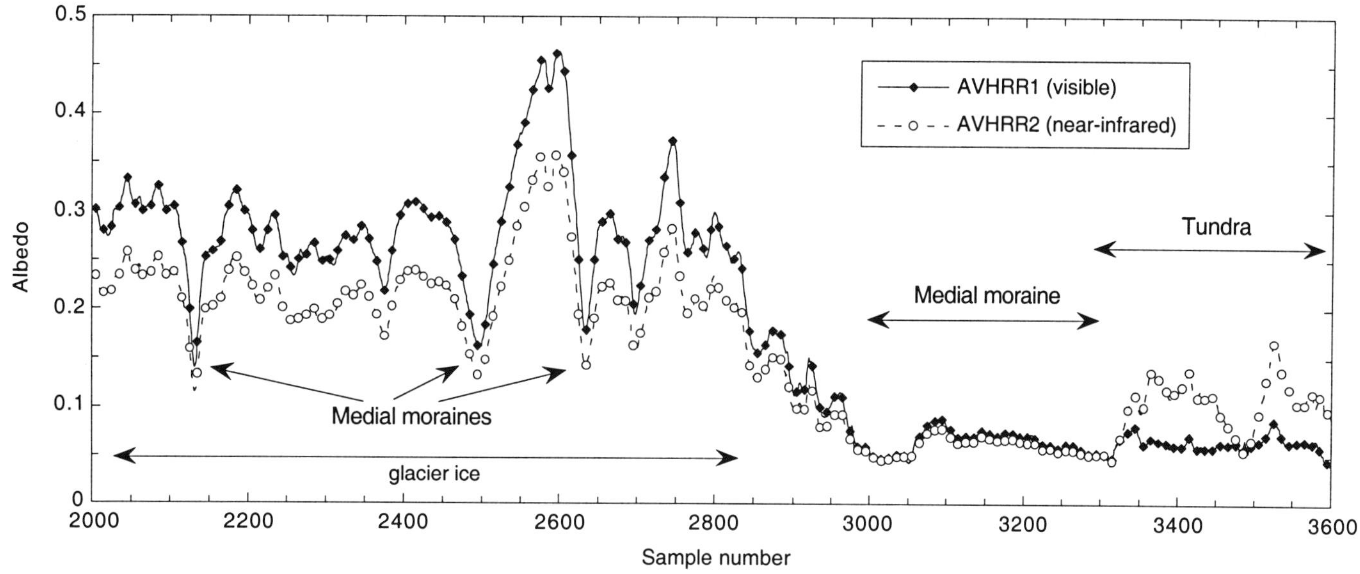

Figure 12.12. Albedo measurements made by a radiometer on board a helicopter flying low (altitude between 10 and 200 m) over Breidamerkurjökull, Iceland. The measurements were made in two NOAA-AVHRR channels (1 and 2). Sample resolution in the field is of the order of a few tens of metres. Every 10th sample is marked on the curves. Surface type is indicated in the figure. Data were kindly provided by W. Greuell, who conducted the experiment.

It is clear that this type of data, also available for Landsat TM channels, is better suited for testing satellite-derived albedos than is the type of data from point measurements at weather stations.

Laser altimetry from planes is another promising development. In combination with accurate GPS, repeated measurements from the same flight line produce data on absolute height changes of the glacier surface. The method is effective because a number of glaciers can be measured in a short time.

The same type of data can also be obtained by using kinematic GPS on a vehicle, for instance on a snow scooter. With differential kinematic GPS, during which one receiver is fixed and the other receiver is moved at a moderate speed, an accuracy of a few centimetres in three-dimensional space can now be achieved [13]. Combining kinematic GPS with mass balance observations and snow radar provides a wealth of information on the dynamics of a glacier. However, such techniques can be used with greater success in accumulation zones than ablation zones, where it is normally more difficult to move around.

12.5 The role of modelling

Observation and modelling should go together. In this book a great deal of attention has been given to models. Working with models is the only way to quantify processes and put these processes into perspective.

Modelling of natural systems like a glacier should be based on physical laws. However, it should be realised that the degree of detail is limited, and that in most cases empirical relations are included to obtain a workable model. Outstanding examples in the context of this book are the treatment of sliding and the calculation of turbulent fluxes.

It is generally not possible to make an absolute quantitative assessment of the state of balance of a natural system, unless it is insensitive to input parameters. However, in such a case the absolute state of balance varies little and is of no interest. High sensitivity in combination with the need to use empirical laws implies that models need to be calibrated before they can have any predictive capabilities.

The mass balance models discussed in Chapter 4 provide a good example. Because of the high sensitivity of glacier mass balance to temperature, the model cannot be used to assess the absolute mean specific balance of a glacier. Even if the models were almost perfect, the uncertainty in the climatological input data would lead to a large error in the mean specific balance. It is a misunderstanding, however, that tuned models would be less reliable for sensitivity experiments. In fact, the reliability of the results of a sensitivity study depend on two main factors: inclusion of the dominant physical processes and careful calibration.

The argument is frequently heard that there is no point in modelling a particular glacier because field data are too scanty. This makes just as much sense as stating that it is useless to perform measurements on a glacier because no model is available. Modelling and making observations should

be combined. Measurements always yield useful constraints for modelling - modelling provides basic information about the dynamic properties of a glacier and may give hints as to what is important to measure.
In the near future models are likely to play an increasingly important role in data assimilation. Traditional mass-balance data from stake readings and pit studies will have to be combined with satellite observations and meteorological data. A great deal can be learned from the techniques developed in numerical weather prediction. Data of different kinds can only be combined in an objective way with a model that describes the essential physics of the system. Therefore we need to improve models and, in particular, to find ways to tie everything together. However, we also need to be more open minded when dealing with 'observations' and 'modelling'.
A measurement of the mean specific balance of a glacier will contain a degree of error. This is due to errors in the single readings at stakes and snow pits, and to the way in which point measurements are converted into a representative number for the entire glacier. In comparing the measured balance for different years, we use observations that were made on different calendar days. This increases the error. As an alternative, we can calculate the mean specific balance with a mass-balance model fed by data from a nearby meteorological station. The result could look like *figure 12.13*. It is traditional to consider the measurements as the 'truth' and the model result as the 'approximation'; but how sure can we be that the measurements are more accurate? One could also argue that a curve that combines the two sources of mass-balance data should yield a better result because, in the end, it makes use of the mass-balance measurements *as well as* meteorological observations. For instance, we can define a best estimate as

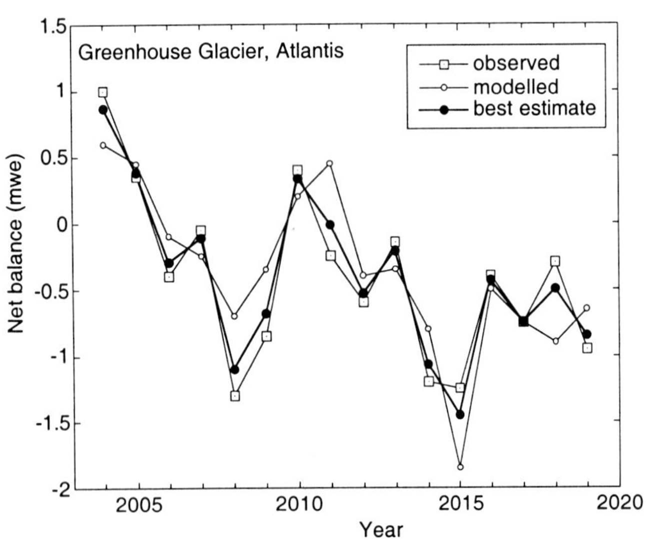

Figure 12.13. Direct observations do not necessarily provide the best estimate of a glacier's mean specific balance. By means of a mass balance model, meteorological input can be used to obtain the best estimate.

$$B_{best} = \frac{\sigma_{met} B_{obs} + \sigma_{obs} B_{met}}{\sigma_{obs} + \sigma_{met}} . \qquad (12.13)$$

Here B_{obs} and B_{met} are the net balances obtained from observation and from modelling with the meteorological input data, respectively. The corresponding error estimates are σ_{obs} and σ_{met}. In the example of *figure 12.13* it was assumed that the value of σ_{met} was twice the value of σ_{obs}.
Clearly, this is a very simple example. In the end we also want to include information from kinematic GPS, laser altimeter, satellite images, etc. in a single model that will then produce a diagnosis of the current state of balance of a glacier. Designing such a sophisticated model will be one of the challenges for glacier research in the coming decade.

References and further reading

1. *Satellite Image Atlas of Glaciers in the World* (eds. R.S. Williams and J.G. Ferrigno). Various volumes, published as Professional Paper No. 1386 of the U.S. Geological Survey.

2. O. Orheim and B.K. Lucchitta (1988): Numerical analysis of Landsat Thematic Mapper images of Antarctica: surface temperatures and physical properties. *Annals of Glaciology* **11**, 109-120.

   R.S. Williams, D.K. Hall and C.S. Benson (1991): Analysis of glacier facies using satellite techniques. *Journal of Glaciology* **37**, 120-128.

   L. Boresjö-Bronge and C. Bronge (1999): Ice and snow-type classification in the Vestfold Hills, East Antarctica, using Landsat-TM data and ground radiometer measurements. *International Journal of Remote Sensing* **20**, 225-240.

3. D.K. Hall, A.T.C. Chang, J.L. Foster, C.S. Benson and W.M. Kovalick (1989): Comparison of in situ and Landsat derived reflectance of Alaskan glaciers. *Remote Sensing of Environment* **28**, 23-31.

   D.K. Hall, R.A. Bindschadler, J.L. Foster, A.T.C. Chang and H Siddalingaiah (1990): Comparison of in situ and TM-derived reflectance of Forbindels Glacier. *Remote Sensing of Environment* **28**, 23-31.

   R. Koelemeijer, J. Oerlemans and S. Tjemkes (1992): The surface reflectance of the Hintereisferner from Landsat 5 TM imagery. *Annals of Glaciology* **17**, 17-22.

   J.-G. Winther (1993): Landsat TM derived and in situ summer reflectance of glaciers in Svalbard. *Polar Research* **12**, 37-55.

D.J. Gratton, P.J. Howart and D.J. Marceau (1993): Using Landsat-5 Thematic Mapper and digital elevation data to determine the net radiation field of a mountain glacier. *Remote Sensing of Environment* **43**, 315-331.

W.H. Knap, B.W. Brock, J. Oerlemans and I.C. Willis (1997): Comparison of Landsat-TM derived and ground-based albedo of Haut Glacier d'Arolla. *International Journal of Remote Sensing* **20**, 3293-310.

C. H. Reijmer, W.H. Knap and J. Oerlemans (1999): The surface albedo of the Vatnajökull Ice Cap, Iceland: a comparison between satellite-derived and ground-based measurements. *Boundary-Layer Meteorology* **92**, 125-144.

4. W.H. Knap, C.H. Reijmer and J. Oerlemans (1998): Narrow band to broadband conversion of Landsat-TM glacier albedos. *International Journal of Remote Sensing* **20**, 2091-2110.

5. B.J. Choudhury and A.T.C. Chang (1981): On the angular variation of solar reflectance of snow. *Journal of Geophysical Research* **86** (C1), 465-472.

J. Dozier, S.R. Schneider and D.F. McGinnis Jr., (1981): Effect of grain size and snowpack water equivalance on visible and near-infrared satellite observations of snow. *Water Resources Research* **17** (4): 1213-1221.

S.G. Warren (1982): Optical properties of snow. *Reviews of Geophysics and Space Physics* **20**, 67-89.

M. Kuhn (1985): Bidirectional reflectance of polar snow and Alpine snow surfaces. *Annals of Glaciology* **6**, 164-167.

J. Dozier (1989): Spectral signature of alpine snow cover from the Landsat Thematic Mapper. *Remote Sensing of Environment* **28**, 9-22.

S.G. Warren, R.E. Brandt and P. O'Rawe Hinton (1998): Effect of surface roughness on bidirectional reflectance of Antarctic snow. *Journal of Geophysical Research* **103** (E11), 25789-25807.

J.-G. Winther, S. Gerland, J.B. Ørbaek, B. Ivanov, A. Blanco and J. Boike (1999): Spectral reflectance of melting snow in a high Arctic watershed on Svalbard: some implications for optical remote sensing studies. *Hydrological Processes* **13**, 2033-2049.

6. W.H. Knap and C.H. Reijmer (1998): Anisotropy of the reflected radiation field over melting glacier ice: measurements in Landsat TM bands 2 and 4. *Remote Sensing of Environment* **65**, 93-104.

W. Greuell and M. de Ruyter de Wildt (1999): Anisotropic reflection by melting glacier ice: measurements and parametrizations in Landsat TM bands 2 and 4. *Remote Sensing of Environment* **70**, 265-277.

7. P. Abel (1990): Prelaunch calibration of the NOAA-11 AVHRR visible and near IR channels. *Remote Sensing of Environment* **31**, 227-229.

G.F. Epema (1990): Determination of planetary reflectance for Landsat-5 Thematic-Mapper tapes processed by Earthnet (Italy). *ESA Journal 1990*, Vol. 14.

8. M.S. Moran, R.D. Jackson, T.R. Clatke, J. Qi, F. Cabot, K.J. Thorne and B.L. Markham (1995): Reflectance factor retrieval from Landsat TM and SPOT HRV data for bright and dark targets. *Remote Sensing of Environment* **52**, 218-230.

9. P. Koepke (1989): Removal of atmospheric effects from AVHRR albedos. *Journal of Applied Meteorology* **28**, 1341-1348.

10. K.C. Partington (1998): Discrimination of glacier facies using multi-temporal SAR data. *Journal of Glaciology* **44** (146), 42-53.

11. R. Gens and J.L. van Genderen (1996): SAR interferometry - issues, techniques, applications. *International Journal of Remote Sensing* **17**, 1803-1835.

D. Massonet and K.L. Feigl (1998): Radar interferometry and its application to changes in the Earth's surface. *Reviews of Geophysics* **36**, 441-500.

D.R. Fatland and C.S. Lingle (1998): Analysis of the 1993-95 Bering Glacier (Alaska) surge using differential SAR interferometry. *Journal of Glaciology* **44** (148), 532-546.

R. Michel and E. Rignot (1999): Flow of Glaciar Moreno, Argentina, from repeat-pass Shuttle Imaging Radar images: comparison of the phase correlation method with radar interferometry. *Journal of Glaciology* **45** (149), 93-100.

R.R. Forster, E. Rignot, B.L. Isacks and K.C. Jezek (1999): Interferometric radar observations of Glaciares Europa and Penguin, Hielo Patagónico Sur, Chile. *Journal of Glaciology* **45** (150), 325-337.

J.A. Dowdeswell, B. Unwin, A.-M. Nuttall and D.J. Wingham (1999): Velocity structure, flow instability and mass flux on a large Arctic ice cap from satellite radar interferometry. *Earth and Planetary Science Letters* **167**, 131-140.

N. Reeh, S.N. Madsen and J.J. Mohr (1999): Combining SAR interferometry and the equation of continuity to estimate the three-dimensional glacier surface-velocity vector. *Journal of Glaciology* **45** (151), 533-538.

12. H. Rott M. Stuefer, A. Siegel, P. Skvarca and A. Eckstaller (1998): Mass fluxes and dynamics of Moreno Glacier, Southern Patagonia Icefield. *Geophysical Research Letters* **25** (9), 1407-1410.

13. T. Eiken, J.O. Hagen and K. Melvold (1997): Kinematic GPS survey of geometry changes on Svalbard glaciers. *Annals of Glaciology* **24**, 157-163.

F.M. Jacobsen and W.H. Theakstone (1997): Monitoring glacier changes using a global positioning system in differential mode. *Annals of Glaciology* **24**, 314-319.